Dyes and Pigments - Novel Applications and Waste Treatment

Edited by Raffaello Papadakis

Published in London, United Kingdom

IntechOpen

Supporting open minds since 2005

Dyes and Pigments - Novel Applications and Waste Treatment
http://dx.doi.org/10.5772/intechopen.90970
Edited by Raffaello Papadakis

Contributors
Mariame Coulibaly, Hassen Agougui, Youssef Guesmi, Mahjoub Jabli, Virendra Kumar Gupta, Deepti Pargai, Mohamed Chiban, Ahmed Zaghloul, Ridouan Benhiti, Rachid Aziam, Abdeljalil Ait Ichou, Mhamed Abali, Amina Soudani, Fouad Sinan, Mohamed Zerbet, Raffaello Papadakis, Ioanna Deligkiozi, Hu Li, Yuanyuan Han, Bo Tian, Haiyan Zhao, Syed Hassan Mujtaba Jafri, Indriana Kartini, Adhi Dwi Hatmanto, Parameswari Kalivel, Ashok Kumar, Utkarsh Dixit, Kaman Singh, Satya Prakash Gupta, Mirza S. Jamal Beg

First published in London, United Kingdom, 2021 by IntechOpen
IntechOpen is the global imprint of INTECHOPEN LIMITED, registered in England and Wales, registration number: 11086078, 5 Princes Gate Court, London, SW7 2QJ, United Kingdom
Printed in Croatia

British Library Cataloguing-in-Publication Data
A catalogue record for this book is available from the British Library

Additional hard and PDF copies can be obtained from orders@intechopen.com

Dyes and Pigments - Novel Applications and Waste Treatment
Edited by Raffaello Papadakis
p. cm.
Print ISBN 978-1-83968-614-6
Online ISBN 978-1-83968-615-3
eBook (PDF) ISBN 978-1-83968-616-0

For EU product safety concerns:
IN TECH d.o.o., Prolaz Marije Krucifikse Kozulić 3, 51000 Rijeka, Croatia, info@intechopen.com or visit our website at intechopen.com.

Meet the editor

Dr. Raffaello Papadakis is a chemical engineer (MEng) with a Ph.D. in Physical Organic Chemistry from the Technical University of Athens. He has worked as a postdoc at the Institute of Molecular Sciences, CNRS/Aix-Marseille University, France, focusing on water oxidation catalysts in the research group of Dr. Thierry Tron and later in the group of Dr. Henrik Ottosson (Uppsala University, Sweden) specializing in the excited state (anti) aromaticity and graphene photochemistry. His research interests revolve around physical organic and materials chemistry (particularly the chemistry of graphene and polysaccharides). He is the author/co-author of thirty peer-reviewed scientific papers and book chapters. Dr. Papadakis currently works as a senior research scientist at TdB Labs, Uppsala, Sweden, being particularly interested in the fluorescent labeling of polysaccharides.

Contents

Preface **XI**

Section 1
Introduction **1**

Chapter 1 **3**
Introductory Chapter: Dyes and Pigments - Past, Present, and Future
by Raffaello Papadakis

Section 2
Novel Applications of Dyes and Pigments **7**

Chapter 2 **9**
Probing Solvation Effects in Binary Solvent Mixtures with the Use of Solvatochromic Dyes
by Ioanna Deligkiozi and Raffaello Papadakis

Chapter 3 **35**
Dyes as Labels in Biosensing
by Hu Li, Yuanyuan Han, Haiyan Zhao, Hassan Jafri and Bo Tian

Chapter 4 **59**
Natural Dyes: From Cotton Fabrics to Solar Cells
by Indriana Kartini and Adhi Dwi Hatmanto

Chapter 5 **81**
Photochromic Dyes for Smart Textiles
by Virendra Kumar Gupta

Chapter 6 **99**
Application of Smart and Functional Dyes in Textiles
by Deepti Pargai

Chapter 7 **117**
Applications of Metal Complexes Dyes in Analytical Chemistry
by Mariame Coulibaly

Chapter 8 **131**
Structure and Properties of Dyes and Pigments
by Ashok Kumar, Utkarsh Dixit, Kaman Singh, Satya Prakash Gupta and Mirza S. Jamal Beg

Section 3
Dye Waste Treatment **151**

Chapter 9 **153**
A Brief Comparative Study on Removal of Toxic Dyes by Different Types of Clay
by Ahmed Zaghloul, Ridouan Benhiti, Rachid Aziam, Abdeljalil Ait Ichou, Mhamed Abali, Amina Soudani, Fouad Sinan, Mohamed Zerbet and Mohamed Chiban

Chapter 10 **167**
Preparation of Functionalized Hydroxyapatite with Biopolymers as Efficient Adsorbents of Methylene Blue
by Hassen Agougui, Youssef Guesmi and Mahjoub Jabli

Chapter 11 **187**
Treatment of Textile Dyeing Waste Water Using TiO_2/Zn Electrode by Spray Pyrolysis in Electrocoagulation Process
by Parameswari Kalivel

Preface

Dyes and pigments play an important role in our everyday lives. Synthetic dyes and pigments are found in numerous objects we utilize every day, from dyes coloring our clothes and pigments in our paintings to advanced dyes used in recordable DVDs and innovative fluorescent dyes employed in live-cell imaging. The versatility of the known dyes and pigments is enormous and their uses and applications are countless. The global demand for new dyes and pigments with novel properties and applications is constantly growing, and thus research attempts to develop these new dyes and pigments are currently intensifying. Of course, a great deal of the currently conducted research aims at a better understanding of the properties and application scope of existing dyes and pigments. However, bulk volumes of dyes and pigments and related byproducts associated with their production and end uses are contaminating the environment and polluting valuable water resources.

Acknowledging these facts, this book covers two main areas of research, and the book is organized into three sections. The first section is the Introduction. In the second section, the book explores new trends in the research of dyes and pigments, focusing on smart and functional dyes and pigments and their uses in textile dying, analytical chemistry, bioimaging and biosensing, and novel chromic and energy-related applications.

The third section examines new trends in the treatment of dye and pigment waste. Techniques employing newly developed materials or modified clays as well as electrochemical methods are described.

As academic editor of this book, I am grateful to all authors who have contributed chapters covering all pertinent research areas that address important scientific, industrial, and technological problems. It has been my pleasure and privilege to read, interact with, and learn from the work of all the authors. A big thanks goes to all of you for the fantastic job you have done.

I am grateful to my old friend and colleague Dr. Dimitris Matiadis (researcher at NCSR Demokritos, Athens, Greece) for the stimulating discussions related to novel synthetic dyes (especially fluorescent ones) and my lovely wife (Dr. Kalliopi Tzavlaki) who always makes sure to keep me updated regarding new trends in bioimaging and bioscience.

Last (but not least), I would like to acknowledge TdB Labs, the company I am happy to have worked at as a Senior Research Scientist since 2019, for providing me with space, time, and motivation to learn more about the intriguing and wide research area of fluorescent labeling and chemistry of synthetic dyes.

Dr. Raffaello Papadakis
TdB Labs,
Uppsala, Sweden

Section 1

Introduction

Chapter 1

Introductory Chapter: Dyes and Pigments - Past, Present, and Future

Raffaello Papadakis

1. Introduction

Dyes and pigments have been playing an undoubtedly important role in human life since the ancient years. Today their mass production is well established, and a vast number of dyes and pigments are globally produced. Currently, the industrial interest for new dyes and pigments with special properties is constantly growing. This has triggered significant research attempts all over the world and new multifunctional dyes and pigments suitable for novel, hi-tech applications have been proposed/created. The steady growth of the global dyes and pigments market signifies a vibrant future for research in the corresponding, wide research field.

2. Historical background

Dyes and pigments are narrowly connected to human culture and different types of them have been used from people since the ancient years in order decorate various types of materials including textiles, ceramics, wood etc. [1] It is well documented that more than 2000 years ago, in ancient China, Egypt, Rome and Greece natural dyes and pigments obtained from plant roots, animals or mineral sources were used. [1, 2] They were mostly used for decorative applications but also as protective layers against wear and corrosion of various objects. [2]

The big revolution in the field occurred in the beginning of 19th century when Sir William Henry Perkin produced the first synthetic organic dye, the so-called *mauveine*, using aniline as a starting compound. [1, 3] *Mauveine* proved to be a suitable dyestuff for various types of textiles predominantly silk, and mass production of the *aniline purple* (the original industrial name of *Mauveine*) commenced. Industrial revolution made the production of many more synthetic dyes feasible, and the expansion of synthetic dyes industry became enormous in later years. In fact, the majority of the currently known classes of dyes and pigments were invented during 19th century. [4]

3. Contemporary trends and the future of dyes and pigments

Today, the classes of dyes and pigments have been enriched with numerous entries. The scope of this field extends to and aims at novel high-tech applications such as laser dyes, [5] dyes and pigments for bio-labelling, [6] intravital microscopy applications, [7] and smart sensing devices responding to various external stimuli [8].

IntechOpen

It comes as no surprise that the global dyes and pigments market for the year 2020 was determined to be as large as $ 32.9 billion with an estimated growing rate of about 5% for the next few years. [9] The largest shares of the global production of dyes and pigments are the ones pertaining to textile-dyes and leather-dyes covering together nearly 75% of the global production. Furthermore, printing inks and paper dyes industry hold important shares.

All the above information indicates that the use of dyes and pigments is well-established in everyday life. They appear in our lives dying objects made out of various materials e.g. plastic, wood, metals, ceramics, leather, textiles etc. (see **Figure 1**) and are either applied by the manufacturers of the products or can be by the end-user in order to protect, modify or decorate variety of objects and materials. The variety of paints depending on use and target-surface is enormous and most of the dyes and pigments are readily available to the user. This indicates a vitally tight connection between production and everyday life. Some examples of dyes and pigments used in everyday life are depicted in **Figure 1**. Their versatility is huge, and their applications are massive.

Yet, science is constantly pushing the limits of the corresponding research field and attempts to obtain novel dyes with special functions are unceasingly being made. (Multi)functional dyes and pigments fall in a wider group of materials which can provide certain types of functionality parallel to their main operational scope. In this sense, a functional dye may for instance not only act as protective and/or decorative layer on a material (main scope), but it could act as an environment-responsive component which could in turn render a new material or device responsive and sensitive to various external stimuli (including light, pressure, heat, environment pH changes, solvent polarity etc.). In the large family of functional dyes (and pigments) fall the so-called chromic compounds/materials. These are compounds or materials which are capable of undergoing (ideally) reversible changes in a way that the user can

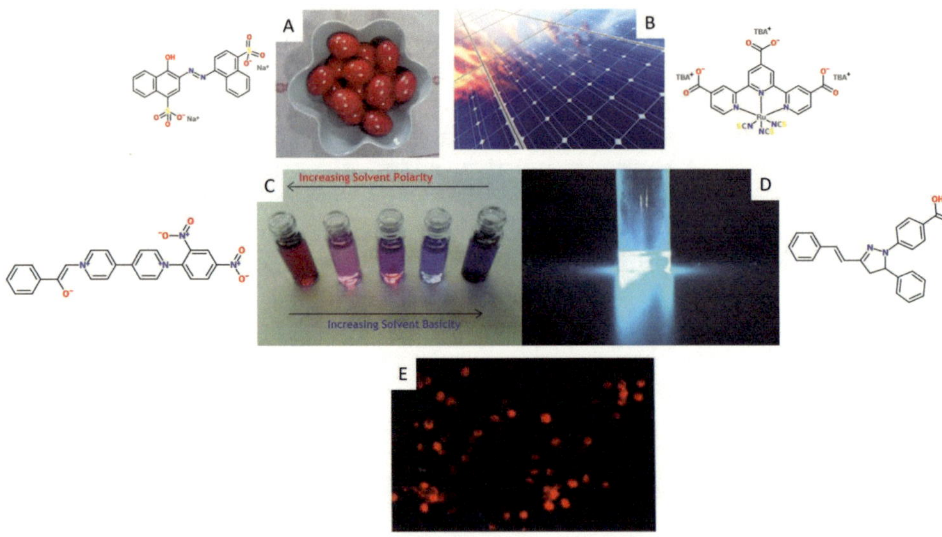

Figure 1.
Examples of dyes used in everyday life as well as in specific applications. (A) Structure of azorubine (also known as E122) a water soluble food-colorant. Photo depicts red-dyed eggs colored with azorubine. (B) Structure of "black-dye" a prominent candidate for dyes-sensitized solar cells. Photograph depicts a solar cell. (C) Structure of a viologen enolate, highly responsive in solvent polarity changes (solvatochromic dye). Photo indicates the drastic color change when moving from water (red) to solvents of lower polarity (adapted with permission by Papadakis et al. [10] (D) a pyrazoline fluorescent dye of high emission intensity. [11] (photo source: author's property). (E) Antonia Red™ Dextran: a novel polysaccharide with fluorescent labelling and its application in cell-imaging (source: TdB Labs) [12].

grasp information regarding the environment of the material/compound. [8] Novel research attempts intent to render sensing systems applicable in microenvironments e.g. living cells and they have led to a sizable family of dyes suitable for microscopy with numerous applications in biology and medicine
(see **Figure 1E**).

4. Conclusion

The variety of commercial dyes and pigments with novel properties and applications is increasing as the needs of the end-users expand. This fact is clearly reflected in the size of the corresponding market and it accounts for the extended global contemporary research endeavors towards new dyes and pigments. Based on these facts, a colorful future is envisioned.

Author details

Raffaello Papadakis
TdB Labs, Uppsala, Sweden

*Address all correspondence to: rafpapadakis@gmail.com

References

[1] Klaus Hunger (Editor) Industrial Dyes Chemistry, Properties, Applications ISBN 3-527-30426-6. WILEY-VCH Verlag GmbH & Co. KGaA, Weinheim, 2003.

[2] Maria J. Melo. History of Natural Dyes in the Ancient Mediterranean World in Handbook of Natural Colorants Edited by Bechtold T. and Mussak R., John Wiley & Sons, Ltd, 2009.

[3] Garfield, S. Mauve: How One Man Invented a Color That Changed the World. W. W. Norton & Company; 2002. ISBN: 978-0393323139.

[4] Gordon P. F., Gregory, P. Organic Chemistry in Colour, Springer-Verlag, Berlin, 1983

[5] Kuehne A. J. C. and Gather M.C. Organic Lasers: Recent Developments on Materials, Device Geometries, and Fabrication Techniques Chem. Rev. 2016, 116 (21), 12823-12864. DOI: 10.1021/acs.chemrev.6b00172

[6] Zhang K. Y., Yu Q., Wei H., Liu S., Zhao Q., and Huang, W. Long-Lived Emissive Probes for Time-Resolved Photoluminescence Bioimaging and Biosensing. Chem. Rev. 2018, 118 (4) 1770-1839 DOI: 10.1021/acs.chemrev.7b00425

[7] Weigert R.(ed.) Advances in Intravital Microscopy. From Basic to Clinical Research. Springer Dordrecht 2014. ISBN 978-94-017-9360-5. DOI 10.1007/978-94-017-9361-2.

[8] Bamfield, P. Chromic Phenomena: Technological Applications of Colour Chemistry: Edn 2nd RSC, Cambridge, 2010. ISBN 978-1-84755-868-8. DOI: 10.1039/9781849731034

[9] Global Dyes & Pigments Market Size Report, 2021-2028 (grandviewresearch.com)

[10] Papadakis R., Deligkiozi I., Tsolomitis A. Spectroscopic investigation of the solvatochromic behavior of a new synthesized non symmetric viologen dye: Study of the solvent–solute interactions. Anal. Bioanal. Chem. 2010, 397, (6), 2253-2259. DOI: 10.1007/s00216-010-3792-7.

[11] Matiadis D., Nowak K., Alexandratou E., Hatzidimitriou A., Sagnou M., Papadakis R. Synthesis and (fluoro)solvatochromism of two 3-styryl-2-pyrazoline derivatives bearing benzoic acid moiety: A spectral, crystallographic and computational study. J. Mol. Liq. 2021, 331, 1, 115737. DOI: 10.1016/j.molliq.2021.115737.

[12] https://tdblabs.se/products/fluorescent-derivatives/antonia-red-products/antonia-red-dextrans/

Section 2

Novel Applications of Dyes and Pigments

Chapter 2

Probing Solvation Effects in Binary Solvent Mixtures with the Use of Solvatochromic Dyes

Ioanna Deligkiozi and Raffaello Papadakis

Abstract

In this work three molecules exhibiting dual sensing solvatochromic behaviors are examined in the context of solvation in binary solvent mixtures (BSMs). The compounds studied involve two functional groups with high responsiveness to solvent polarity namely pentacyanoferrate(II) (PC) and azo groups. Two of these compounds are [2]rotaxanes involving *alpha-* or *beta-* cyclodextrin (CyD) and the third is their CyD-free precursor. The dual solvatochromic behavior of these compounds is investigated in water/ethlylene glycol (EG) mixtures and their dual solvatochromic responses are assessed in terms of the intensity of solvatochromism and the extent of preferential solvation. To achieve this the linear solvation model by Kamlet, Abboud and Taft [*J. Organomet. Chem.* 1983, *48*, 2877–2887] and the two-phase model of solvation by Bagchi and coworkers [*J. Phys. Chem.* 1991, *95*, 3311–3314] are employed. The influence of the presence or lack of CyD (*alpha-* or *beta-*) on these dual solvatochromic sensors is analyzed.

Keywords: solvatochromic dyes, rotaxanes, preferential solvation, (non) specific solute-solvent effects, azo dyes

1. Introduction

Nowadays, solvatochromic probes (SPs) are regularly utilized in various types of applications which require sensing of environmental/medium effects in either a qualitative or a quantitative manner [1–9]. Today there is a large variety of published solvatochromic dyes corresponding to different media, e.g. organic solvents [8, 10], ionic liquids [10, 11], solvent mixtures [10, 12–15] or solvent comprising polarity modifiers [16]. What often appears to be challenging is the choice of a suitable solvatochromic probe for the description of a physicochemical problem encompassing solvent polarity effects. It has been observed that for the same solvent/cosolvent mixture, different solvatochromic dyes may provide different quantitative results [17, 18]. Indeed in many cases, different spectroscopic techniques applied on the same ternary system solvent/cosolvent/probe(solute) may provide different results. Therefore, probing solvent polarity effects and preferential solvation (PS) phenomena occurring in solvent mixtures of two or more

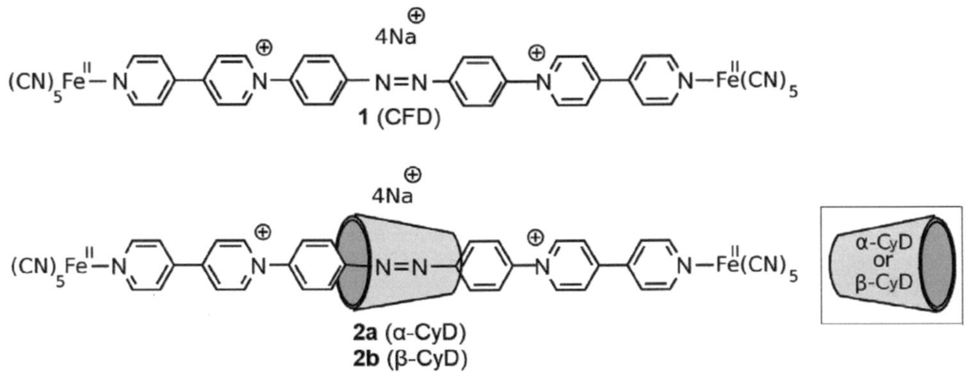

Figure 1.
The three solvatochromic compounds involved in this study.

solvents are considered as highly difficult tasks [17]. The complexity of those physicochemical problems is high and the interpretation of the solvent-solvent or solute-solvent effects sensed by SPs needs to be carefully undertaken. In this work, the authors examine the solvatochromic responses of two probing groups: the azo-group and the pentacyanoferrate(II) group of three molecules dissolved in binary solvent mixtures (BSMs) involving water and ethylene glycol (EG). From the three molecules employed two are [2]rotaxanes involving *alpha-* or *beta-* cyclodextrin (CyD) (compounds **2a** and **2b** respectively, **Figure 1**) and the third is their precursor lacking a CyD wheel (compound **1**, **Figure 1**; in the text will be called cyclodextrin-free dumbbell-like compound or CFD). All three are recently developed solvatochromic compounds [19] and they involve the same π-conjugated viologen-based linear skeletons bearing an azobenzene bridge and pentacyanoferrate(II) end-groups (**Figure 1**). The aforementioned compounds fall under the umbrella of an important family of multifunctional dyes primarily because of the high technological and industrial importance of azobenzene dyes [20–22] as well as the pronounced chromic and redox behavior [23], photochromism [24], photoconductivity [25] and strong electron withdrawing aptitude of viologens (also known as paraquats) [23, 26]. This strong electron accepting capacity of para- and monoquats is vital for the development of push-pull systems [27–30]. Towards the latter milestone the use of suitable electron donating substituents is vital. Papadakis et al. has shown that pentacyanoferrate(II) units can trigger an intense solvatochromic behavior in such systems in various types of media [14, 16, 30, 31] and more recently Deligkiozi et al. hinted that the $n \rightarrow \pi^*$ transitions of the azo group in **1** and **2a,b** are sensitive to solvent polarity [19]. In this work the dual solvatochromic sensing of these compounds is thoroughly examined in terms of solvent-solute, solvent-solvent and preferential solvation effects. In this work water/ethylene glycol mixtures were chosen as perfect BSM candidate-models of dipolar media with substantial contributions in the development of specific effects (mainly H-bonding).

2. Materials and methods

2.1 General information

All correlations (single or multi- parameter linear, polynomial regressions and contribution analyses were carried out using statistical software *R* (ver.:3.5.3). Integrations as well as the graphical determination of isosolvation points were all performed using QtiPlot (ver.: 0.9.9).

All compounds involved in this work (**1** and **2a,b**; **Figure 1**) have been reported in earlier publication by the author and coworkers and their synthesis, isolation and spectral analysis have been also thoroughly described [19]. All solvatochromic UV–Vis shifts have been recorded on a Perkin-Elmer Lambda 25 UV/Vis spectrophotometer. The deconvolutions of all UV–Vis spectra were implemented according to previous work [19].

2.2 Examining the stability of compounds 1,2a,b in solution

All solvatochromic compounds used in this work are isolated as stable solid compounds of green-blue color. The measurements presented were conducted in fresh solutions of each compound in the desired H_2O/EG mixtures (typically prepared 15 min prior to measurement). That time corresponds to the equilibration time (each sample was vigorously stirred after mixing). Directly after this period of time their electronic absorption spectra were recorded. It was observed that in all cases the solutions remained unmodified as concluded through check of the absorbances of the bands maxima which were found to be stable for at least 30 minutes after equilibration. This observation clearly indicates that all three compounds are stable in solution and therefore suitable for the current investigation.

2.3 Preferential solvation model

In this work a renowned PS model is employed in order to describe PS phenomena occurring in BSMs comprising solvatochromic solutes. The model was introduced by Bagchi and coworkers about thirty years ago and is also known as "the two-phase model of solvation" (TPMS) [32–34]. TPMS considers that solvent molecules in a BSM are distributed between a local phase and a bulk phase according to Eq. 1. The local phase lies in the vicinity of the solvation area.

$$\overline{S_2} + S_1 \rightleftharpoons \overline{S_1} + S_2 \tag{1}$$

In Eq. 1 S_1 and S_2 symbolize the two mixed solvents in the bulk phase while $\overline{S_1}$ and $\overline{S_2}$ symbolize the two solvents in the solvation (local) phase. Throughout this work water will be considered as S_1 whereas EG as S_2. At the equilibrium described by Eq. 1, PS constant (K_{ps}) will be related to an expression comprising both solvent-solvent solute-solvent interaction energies (Eq. 2).

$$kTlnK_{PS} = [\epsilon_{S2} - \epsilon_{S1}] + \left[(N_1 - N_2)\epsilon_{12} - (N_1^0 - N_2^0)\epsilon_{12}^0 - N_1\epsilon_{11} + N_1^0\epsilon_{11}^0 + \right. \\ \left. + N_2\epsilon_{22} - N_2^0\epsilon_{22}^0 + \frac{(\epsilon_{11} - \epsilon_{22})}{2} - \frac{(\epsilon_{11}^0 - \epsilon_{22}^0)}{2}\right] \tag{2}$$

In Eq. 2 ϵ_{Si} are the interaction energies among the solute and i-solvent while ϵ_{ij} corresponds to the interaction energies between solvents i and j. N_i corresponds to the number of i-solvent molecules. The bulk phase in solvent molecule numbers is designated with the superscript 0. It is noteworthy that Eq. 2 involves two terms. The first bracketed term in the right-hand side of the equation corresponds to the contribution of solute-solvent interactions, while the terms in the second bracket describe the solvent non-ideality effects.

Finally, Eq. 3, provides K_{ps} (the preferential solvation constant) related to both bulk (x) and local (y) solvent mole fractions along with the measured transition energies of the indicator solute (E_T) observed in neat solvent S_1 ($E_{T,1}$), neat solvent S_2 ($E_{T,2}$) and the mixture of S_1 and S_2 ($E_{T,m}$).

$$K_{PS} = \frac{y_1 x_2}{y_2 x_1} = \frac{E_{T,m} - E_{T,2}}{E_{T,1} - E_{T,2}} \bullet \frac{x_2}{x_1} \tag{3}$$

2.4 Applying the CNIBS/R-K equation

In order to determine isosolvation points (see below) pertaining to PS occurring in solutions of **1** and **2a,b** in aqueous EG, Redlich–Kister (CNIBS/R–K) equation [35], was employed so as to algebraically describe the dependence of experimental transition energy E_T ($n \rightarrow \pi^*$ (azo) and Metal to Ligand Charge Transfer (MLCT)) on solvent/cosolvent bulk mole fractions (x_1, x_2) (Eq. (4)). In this work water is considered as solvent S_1 (i.e. water mole fraction is x_1). Noteworthy, Eq. 4 yields E_T values corresponding to neat solvents S_1 and S_2 ($E^0_{T,1}$ *and* $E^0_{T,2}$) when any of x_2 and x_1 is set to 0 respectively.

$$E_{T,m} = x_1 E^0_{T,1} + x_2 E^0_{T,2} + x_1 x_2 \sum_{j=0}^{k} A_j (x_1 - x_2)^j \tag{4}$$

2.5 Determining isosolvation points

For a BSM involving solvents S_1 and S_2 and a solvatochromic solute, when $x^T_{iso}(S_2) < 0.5$ then PS of the solute by solvent S_2 is observed and vice versa when $x^T_{iso}(S_1) > 0.5$ [18] (where T corresponds to the transition of interest i.e. MLCT (PCF) or $n \rightarrow \pi^*$ (azo) for this study).

$$x^T_{iso} = x^T \text{ at which } E_{T,m} = \frac{E^0_{T,1} + E^0_{T,2}}{2.}$$

For the determination of x^T_{iso} the polynomial expressions obtained through Eq. 5 were utilized. $E_{T,m} = f(x_1)$ were first plotted and then x^T_{iso} were determined graphically using the data reader tool of Qtiplot 0.9.9.

2.6 Quantifying the difference in the extent of PS

In order to quantify the difference of the extent of PS in aqueous EG through the two different types of transitions (MLCT(PCF) or $n \rightarrow \pi^*$ (azo)) of **1,2a,b** the following integrals difference (**Δ** ∫) was employed.

$$\boldsymbol{\Delta} \int = \left(\int_0^1 y_{EG}^{n\rightarrow\pi^*}(x_{EG})\, dx_{EG} \right) - \left(\int_0^1 y_{EG}^{mlct}(x_{EG})\, dx_{EG} \right) \tag{5}$$

In Eq. 5$y_{EG}^{n\rightarrow\pi^*}$ and y_{EG}^{MLCT} are the local EG molar fractions determined through the TPMS methodology peraining to the $n \rightarrow \pi^*$ (azo) and MLCT (PCF) transitions respectively.

2.7 Single parameter regression analyses

To understand the role of various solvatochromic parameters expressing solvent polarity, single regression analyses were implemented (general Eq. (6)). SSP-LSERs

$$E_{T,m} = E^0_{T,m} + p_i \cdot SP_i, r^2 \quad (6)$$

Where $T = MLCT \; or \; n \rightarrow \pi^*$ and SP a solvent polarity parameter (in this work: E_T^N, π^*, α, or β). r^2: correlation coefficient.

2.8 KAT equation and contribution analysis

Moreover a multiparametric model was employed in order to assess the relative contribution of various solvatochromic parameters expressing solvent polarity, simultaneously. That model is the LSER introduced by Kamlet Abboud and Taft (KAT equation). This renowned LSER (Eq. 7) can provide information on the importance of dipolarity/polarizability, Hydrogen bond donor (HBD) acidity and Hydrogen bond acceptor (HBA) basicity of neat solvents or solvent mixtures.

$$E_{T,m} = E^0_{T,m} + s\pi^* + a\alpha + b\beta \quad (7)$$

Where

$$T = MLCT \; or \; n \rightarrow \pi^*$$

Through Eqs. 8–9 it is possible to determine the relative contribution (r_{s_i}) of each of the involved parameters. The procedure has thoroughly been described in previous works [28, 31, 36].

$$r_{s_i} = \frac{\sigma'_i}{\sum_{i=1}^{n} \sigma'_i} \quad (8)$$

and

	$E_{n\rightarrow\pi^*}$ $(kcal.mol^{-1})^a$			E_{MLCT} $(kcal.mol^{-1})^a$		
x_{EG}	1	2a	2b	1	2a	2b
0	73.12	74.26	74.26	51.46	49.02	50.71
0.051	71.48	73.88	73.50	49.06	48.67	48.85
0.097	72.38	73.50	73.88	48.44	47.16	47.18
0.139	70.77	72.38	72.57	47.26	46.15	47.18
0.178	69.91	71.48	71.48	46.23	45.36	46.67
0.212	69.73	71.48	71.48	46.34	44.50	46.50
0.245	69.73	71.48	71.84	45.29	44.10	46.50
0.392	69.73	70.95	71.66	45.78	43.74	44.85
0.492	69.73	70.77	71.12	45.06	42.94	43.66
0.659	69.73	70.77	70.77	43.95	41.85	42.83
0.854	69.73	70.77	70.77	42.94	41.08	41.94
1	69.73	70.42	70.42	41.60	40.35	40.94

[a]*Data from reference: [19].*

Table 1.
Solvatochromic shifts of **1** *and* **2a,b** *in aqueous EG and solvent polarity parameters.*

$$\sigma_i' = |\sigma_i| \sqrt{\frac{\sum_{j=1}^{m} \left(S_{ij} - \overline{S_i}\right)^2}{\sum_{j=1}^{m} \left(E_{Tj} - \overline{E_T}\right)^2}} \tag{9}$$

where S_i corresponds to the correlation to each of any of the three parameter involved in Eq. 8 for the various solvent/cosolvent molar ratios examined (number of different mole ratios examined in this work: $m = 12$; see **Table 1**).

3. Results and discussion

3.1 The bisensing solvatochromic compounds 1 and 2a,b

Recently Deligkiozi et al. [19] and subsequently Papadakis et al. [37] reported on the solvatochromic behavior of compounds **1, 2a-b** (**Figure 1**). It has been pointed out that the energy of the MLCT transition in these compounds is intensely dependent on the polarity of the environment and its nature and characteristics have been thoroughly described in a series of research works [14–16, 19, 30, 31]. These compounds have been studied in a rather narrow solvent polarity range and specifically in aqueous EG mixtures as well as in neat water and EG. All three compounds are very soluble in both those solvents and their mixtures. Despite the small solvent polarity difference observed when moving form water to EG, the recorded difference in MLCT energy of **1** and **2a,b** was reported to be significantly high, following the sequence:

$$\left|\Delta_{H_2O}^{EG}(\tilde{\nu}_{MLCT}\mathbf{1})\right| = 3451\ cm^{-1} > \left|\Delta_{H_2O}^{EG}(\tilde{\nu}_{MLCT}\mathbf{2a})\right| = 3419\ cm^{-1} >$$

$$> \left|\Delta_{H_2O}^{EG}(\tilde{\nu}_{MLCT}\mathbf{2b})\right| = 3033\ cm^{-1}.$$

Interestingly, all three compounds also exhibit another transition which is significantly influenced by solvent polarity. The latter is attributed to the azobenzene group and corresponds to the forbidden $n \rightarrow \pi^*$ transition of the lone pairs of electrons of the azo nitrogen atoms and is located at λ ranging within 385–410 nm strongly depending on the polarity of the solvent. The solvatochromism of azobenzene-based compounds has been thoroughly investigated in the past and there is clear evidence of the solvent dependent nature of the $n \rightarrow \pi^*$ and $\pi \rightarrow \pi^*$ azo transitions [38–41]. This comes as no surprise as the nitrogen atoms of the azo group can readily interact with solvent molecules (in case of compound **1**) or the interior groups of CyDs (*alpha-* for **2a** and *beta-* for **2b**). Typically, the energy of the $n \rightarrow \pi^*$ transition shifts about 6 nm hypsochromically just upon insertion of the CyD wheel. For instance while in neat water $\lambda_{(n \rightarrow \pi^*)} \cong$ 391 nm for the CFD compound (**1**) while $\lambda_{(n \rightarrow \pi^*)} \cong$ 385 nm when the CyD wheel is threaded and stationed around the azobenzene group (same value for both **2a** and **2b**). Comparable shifts are observed at various mole fractions of water in aqueous EG (see **Table 1**). Yet, the effect of solvents is much more important as shifts of even 25 nm are observed when simply moving from water to EG i.e. two solvents with many similarities when it comes to solvent polarity and structuredness [18]. The observed shifts recorded followed the trend:

$$\left|\Delta_{H_2O}^{EG}(\tilde{\nu}_{n \rightarrow \pi^*}\mathbf{1})\right| = 1186\ cm^{-1} < \left|\Delta_{H_2O}^{EG}(\tilde{\nu}_{n \rightarrow \pi^*}\mathbf{2a})\right| = \left|\Delta_{H_2O}^{EG}(\tilde{\nu}_{n \rightarrow \pi^*}\mathbf{2b})\right| = 1345\ cm^{-1}.$$

It is important to note here that the $n \rightarrow \pi^*$ transition is convoluted with the MLCT and $\pi \rightarrow \pi^*$ transitions and a thorough deconvolution analysis has been already published recently [19]. In this work the used values of the energies for $n \rightarrow \pi^*$ and MLCT transitions for **1** and **2a,b** correspond to the aforementioned published deconvolution values (see **Figure 2** and **Table 1**).

Taken together, there is clear evidence that all three compounds are considered to be bisensing as they involve two functional groups (FC and azo groups) both responding to solvent polarity changes however at different extents (**Table 1**) as it will be thoroughly analyzed.

3.2 Resonance structures of compounds 1 and 2a,b

For a better understanding of the dual solvatochromic behavior of all compounds the analysis of their resonance structures is vital. While resonance structure **I** (**Figure 3**) is more important in the electronic ground state and comprises Fe(II) and the all-aromatic structure of the ligand (L), Resonance structure **II** (**Figure 3**) becomes more important in the MLCT excited state of molecules **1** and **2a,b**. The latter resonance structure comprises the oxidized metal center (Fe(III)) and the quinoidal structured ligand as a result of the acceptance of an electron transferred by Fe(II) upon oxidation occurring via absorption of light (MLCT). (Structure **II** is one of the corresponding resonance structures of the type: $[(\mathbf{Fe^{III}} - \mathbf{L^{+\bullet}})]$.

Finally, structure **III** (**Figure 3**) retains the oxidized Fe^{III} center however displays the possibility of stabilization of the $>N^{+\bullet}$ by the azo group. (It is noteworthy that the azo group is known to stabilize carbocations in a similar fashion [42]). The interplay between Resonance structures **II** and **III** can be alternatively written as: $[(\mathbf{Fe^{III}}-(\boldsymbol{py^{+\bullet}}) - \boldsymbol{azo}] \leftrightarrow [(\mathbf{Fe^{III}}-(\boldsymbol{py^{+}}) - \boldsymbol{azo^{\bullet}}]$.

The latter interaction of the azo group with its partly reduced neighboring viologen pyridin heterocycle obviously influences the $n \rightarrow \pi^*$ transition of the azo group which is in turn largely influenced by its interactions in solution with solvent molecules (this applies only to the case of compound **1**) or the interactions with the CyD interior environment (this is obviously valid only for rotaxanes **2a** and **b**). In any case, through Structures **II** and **III** it becomes obvious that the $n \rightarrow \pi^*$ transitions and the solvatochromism of the azo group is largely influenced by the $(Fe^{II} - L)$ system attached to it in π-conjugation.

Frontier orbital representations of the tetraanion of dye **1** ($\mathbf{1}^{4-}$) (for more details see ref.: [19]) also support the fact that the FC groups and azo groups are behaving as electron donating since the HOMO are mostly localized in the regions of the FC and azobenzene moieties (**Figure 4**). On the other hand, LUMO are mostly localized around the quaternized electron deficient viologen parts of the solvatochromic compounds (**Figure 4**). This is an additional hint to the dominating resonance structures presented in **Figure 3**.

3.3 Solute-solvent interactions

It is also noteworthy that in cases **2a,b** the $n \rightarrow \pi^*$ transition of the azo group is even more affected by the behavior of the 4,4′-bipyridine-Fe(II) system (which is in π-conjugation with the azobenzene moiety) than happens in case **1**. This is associated with the fact that CyD (either *alpha-* or *beta-*) does not allow any direct interaction of the azo group with any of the solvents (water or EG; see **Figure 5**).

It is known that solvents with $logP_{oc/w} < -0.3$ cannot penetrate the highly lipophilic cavity of cyclodextrins (for EG it is $logP_{oc/w} = -1.3 < -0.3$ rendering it very hydrophilic to enter the CyD cavity ($P_{oc/w}$ is the 1-octanol/water partition coefficient) [43]. This is clearly manifested by the linear correlation between the

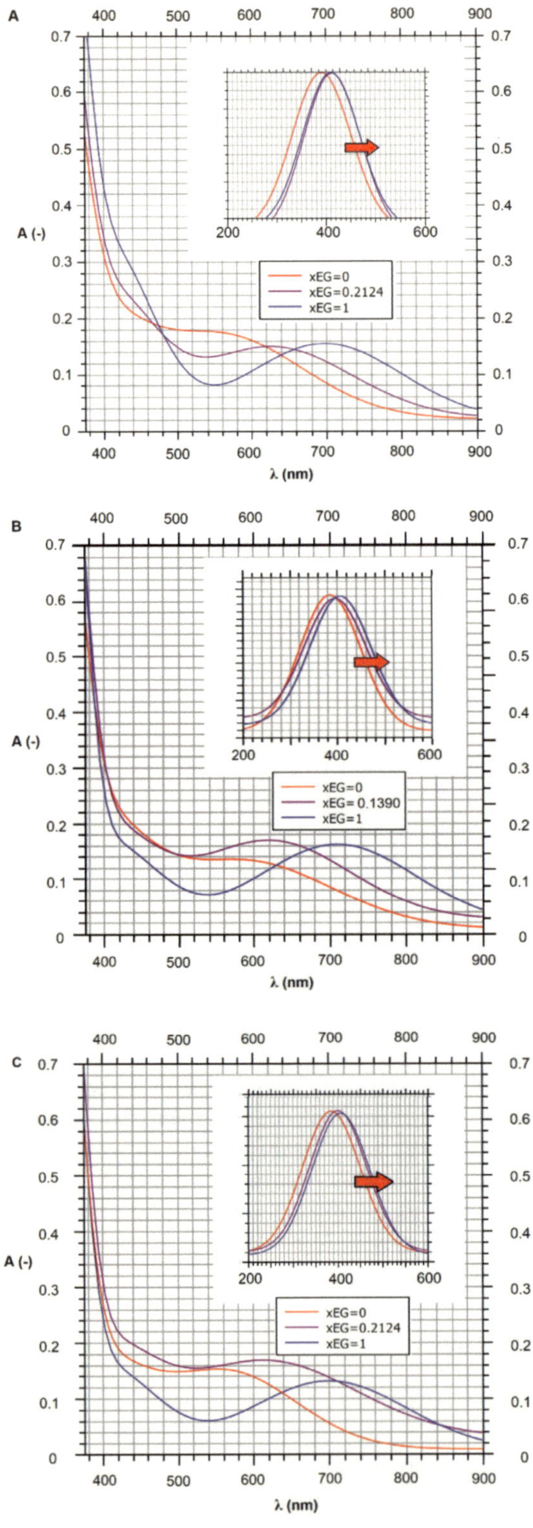

Figure 2.
The electronic spectra of compounds A) **1**, *B)* **2a**, *and C)* **2b** *recorded in water, EG and selected water/EG mixtures. Displayed wavelength range 380–900 nm. Insets indicate the* n→π* *bands after deconvolution.*

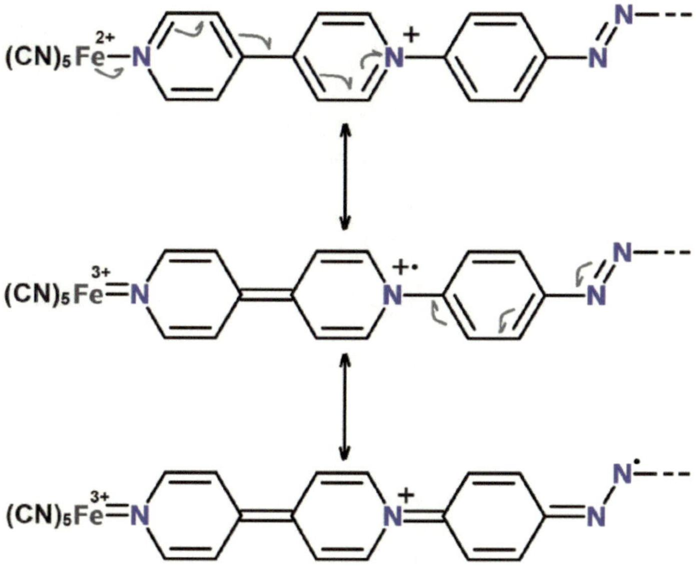

Figure 3.
Characteristic resonance structures of compound 1.

MLCT and $n{\rightarrow}\pi^*$ transitions energies of **2a,b** at various mole fractions of EG (see **Figure 6**). In case of compound **1** such a linear behavior is not observed (**Figure 6**).

3.4 Quantification of the solvatochromism of the FC and azo groups

A pertinent way to quantify, predict and rationalize solvent effects is the use of linear solvation energy relationships (LSERs). This approach has been employed in numerous research works focusing on solvent effect on a large variety of physico-chemical properties [8, 10, 44–46]. The solvatochromism of PC complexes has been thoroughly investigated in this fashion as well [15, 16, 19, 31]. Particularly in case of PC complexes bearing pyridinic ligands (such as **1** and **2a,b**) it has been shown that the MLCT transition energies are largely affected by the dipolarity/polarizability of the medium as well as hydrogen bond donor (HBD) and Lewis acidity [15, 16, 19, 31]. Deligkiozi et al. recently reported the corresponding correlations and furthermore hinted that the energy of the $n \rightarrow \pi^*$ of the azo group in **1** and **2a,b** is also sensitive to solvent polarity changes, thus revealing a dual sensing aptitude of solvent polarity for these compounds [19]. Nevertheless, that work mainly focused on the MLCT transitions of these three compounds. Herein, the author focuses further on the solvents effects on the solvatochromic behavior of the azo group of **1** and **2a,b** and examines this dual solvatochromic behavior.

Plots $E_{n\rightarrow\pi^*}\ vs E_{MLCT}$ (**Figure 6**) indicate fairly good linear correlations for the two rotaxanes (**2a,b**) however a severe deviation from linearity in case of the CFD compound (**1**). This is a stimulating finding pertaining to the structural diversity between the two rotaxanes and their CFD precursor. What this finding implies is that for compounds **2a,b** the medium responsive behavior of the azo group (expressed through $n \rightarrow \pi^*$ transitions) is expected to be analogous to that of the FC groups, at least qualitatively. Furthermore, the duality of solvent polarity sensing aptitude of **1** is anticipated as more pronounced. In order to shed light on these two hypotheses a series of correlations utilizing monoparametric LSERs was accomplished for all three compounds.

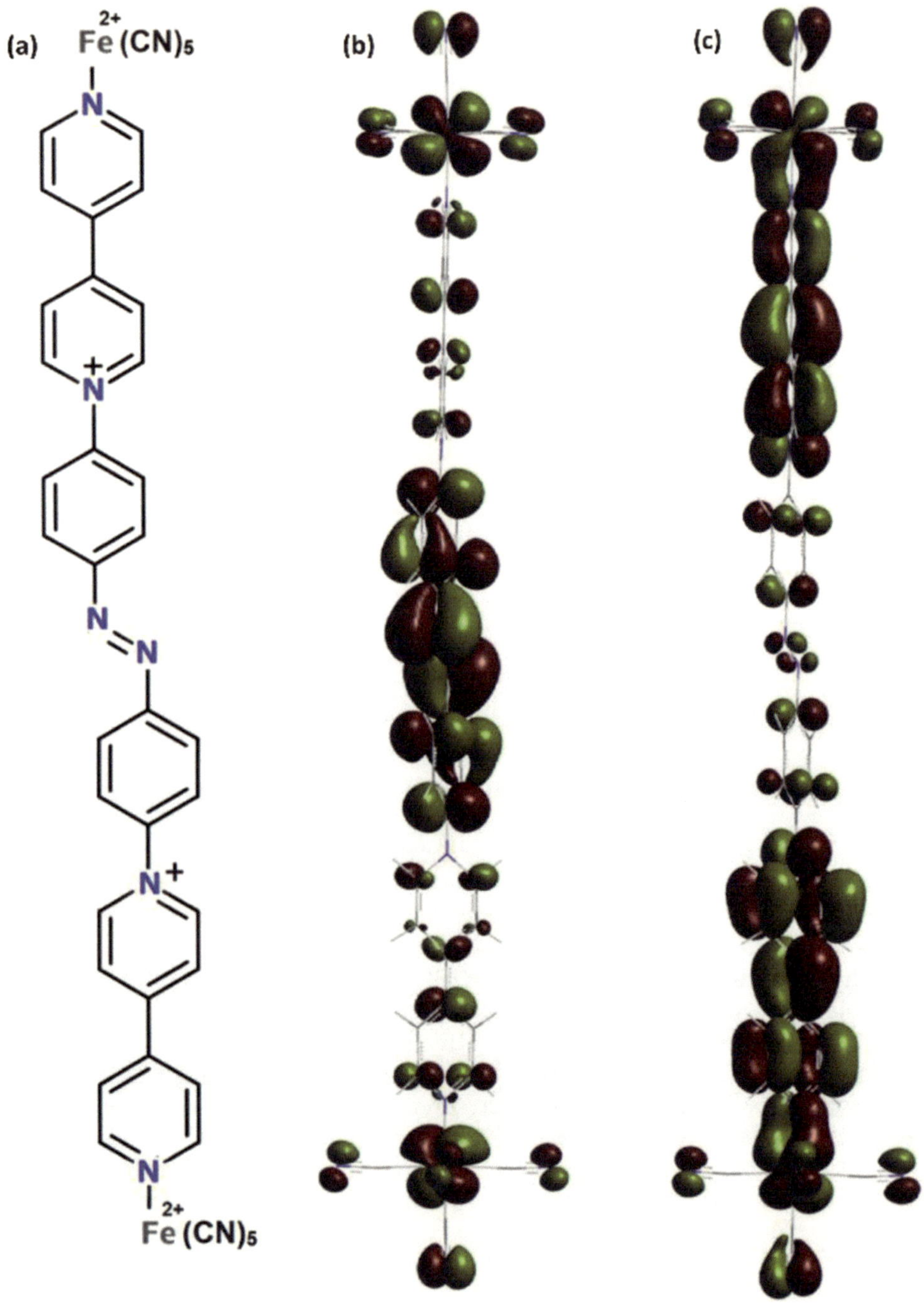

Figure 4.
Structure of the anion of dye 1 (a) and illustrations of the frontier molecular orbitals (MOs) of the anion of the anion if dye 1^{4-}*: HOMO (b) and LUMO (c) calculated on the B3P86/6–311++G(d,p) basis/vacuum (0.02 contour plots).*

3.5 Single solvent polarity parameter involving LSERs

Single solvent polarity parameter involving LSERs (SSP-LSERs) were employed in order to investigate the importance of various solvent polarity parameters on $n \rightarrow \pi^*$ and MLCT transitions of **1** and **2a,b**. The SPPs utilized were the following:

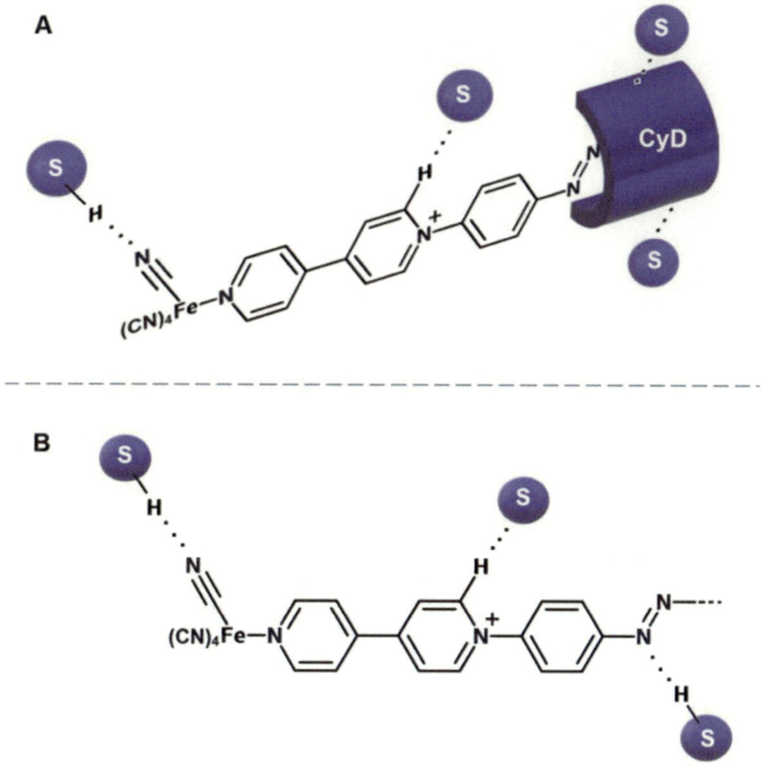

Figure 5.
*Illustration of the possible solute-solvent interactions in A) the rotaxanes **2a,b** and B) their CFD precursor **1** (S represents a solvent molecule or a solvent-cosolvent complex).*

Reichardt's solvent polarity scale: E_T^N, π^*, α, and β. The latter three are parameters involved in KAT equation expressing dipolarity/polarizability, HBD-acidity, and HBA-basicity respectively [47]. It is obvious that rotaxanes **2a** and **2b** exhibit $n \rightarrow \pi^*$ transition energies which correlate linearly with Reichardt's polarity scale E_T^N varying within 1.000 for water and 0.790 for EG (**Figure** 7). In contrast, the fitted curve between $n \rightarrow \pi^*$ transition energies of **1** and E_T^N is not a straight line (**Figure** 7). This finding for the solvatochromism of the azo group of compound **1** implies a different behavior compared to either of the rotaxanes **2a,b** or even the MLCT transitions of the same compound. Moreover, as E_T^N is a measure of dipolarity and Lewis acidity of the medium, the aforementioned results indicate a significantly lower dependence of the $n \rightarrow \pi^*$ energies on these solvent polarity features for compound **1**. Similar results were obtained for the correlations between $n \rightarrow \pi^*$ transitions for all three compounds and parameters π* and α, expressing solvent dipolarity/polarizability and HBD-acidity respectively (**Figures** 7 and **8**). In those cases compound **1** continued to differ from compounds **2a-b.** This comes as no surprise as the connection between E_T^N scale and KAT parameters π* and α is known (as already mentioned E_T^N is a measure of solvent dipolarity and Lewis acidity) [10]. Very interestingly, the $n \rightarrow \pi^*$ energies of compound **1** correlate better with parameter β (involved in KAT equation and expressing HBA-basicity) than happens in case of the two rotaxanes **2a,b** (**Figure** 8).

For all three compounds the two main solvatochromic functional groups are the PCF and azo groups which both behave as fairly good HBA-bases being prone to formation of hydrogen bonds between the –CN and –N=N– groups respectively and

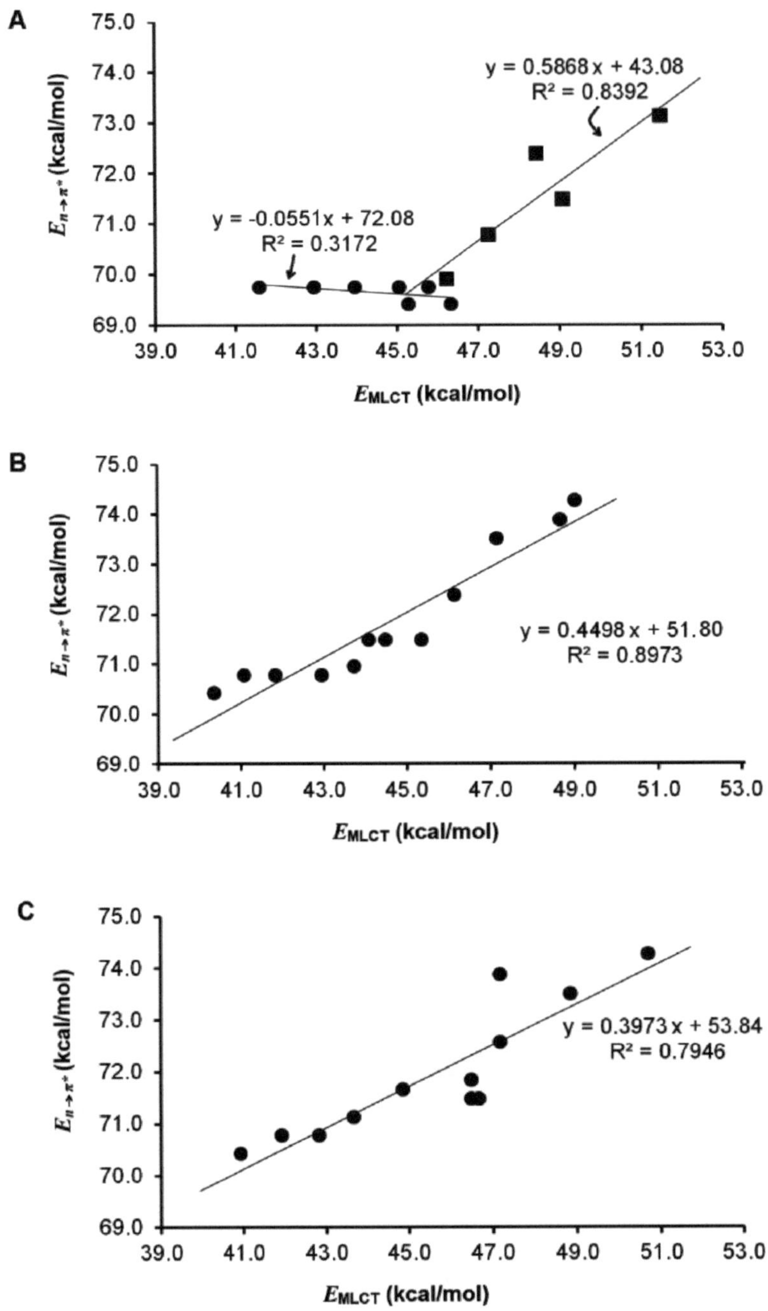

Figure 6.
Plots of the type $E_{n\rightarrow\pi^*}$ vsE_{MLCT} *for (A) compound* **1**, *(B) compound* **2a**, *and (C) compound* **2b**.

hydrogen atoms of protic solvents (like water and EG). Therefore, the nearly linear correlation observed between $E_{n\rightarrow\pi^*}$ and parameter β for compound **1**, indicates another type of solute-solvent interaction less pronounced in rotaxanes **2a,b** (**Figure 5**). The *ortho*-hydrogen atoms of pyridinium salts lying very close to the quaternized nitrogen are known to undergo deuterium-exchange [49–51]. The *ortho*-protons are significantly deshielded with chemical shifts around 9 ppm and sometimes close to 10 ppm (9.30 for **1**, 9.28 for **2b** and 9.27 ppm for **2a**) [19]. It is therefore anticipated that these hydrogen atoms are prone to interactions with polar

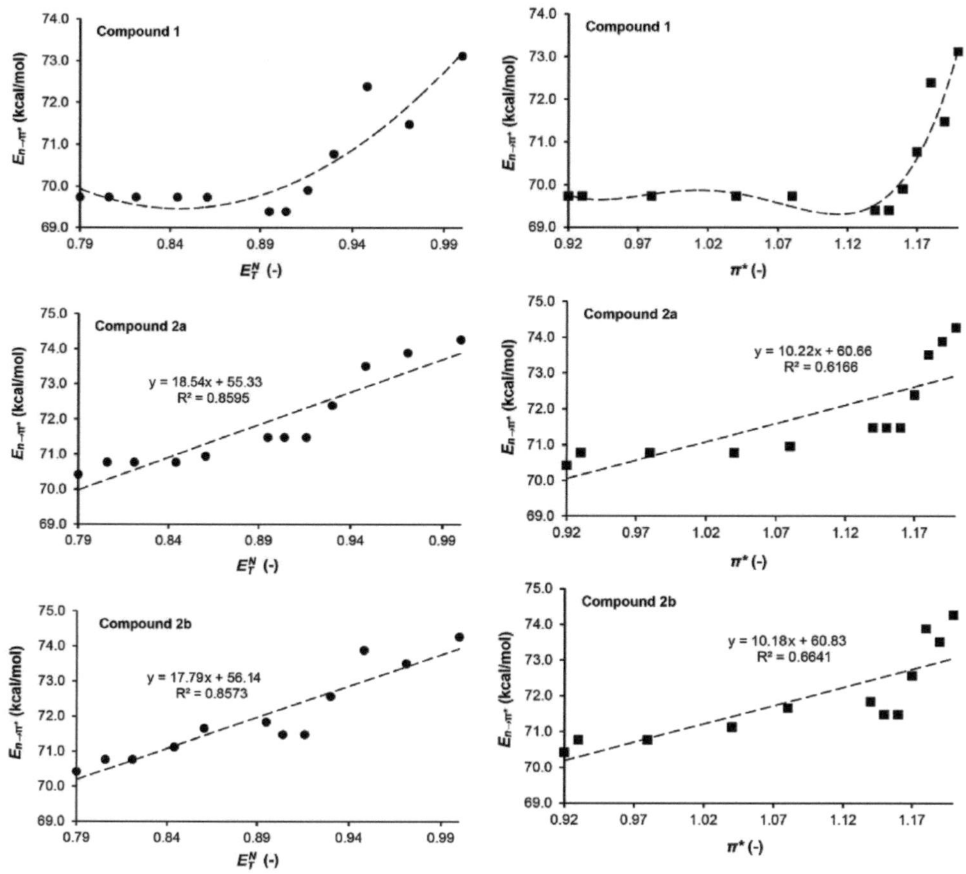

Figure 7.
Left column: *Plots of $n \rightarrow \pi^*$ energies determined in aqueous EG mixtures* vs *Reichardt's solvent polarity scale for compounds **1**, **2a** and **2b**. Right column: Plots of $n \rightarrow \pi^*$ energies determined in aqueous EG mixtures* vs *solvent polarity parameter π^* for compounds **1**, **2a** and **2b** [48].*

solvent molecules. Nevertheless, due to the presence of *alpha-* or *beta-* CyD in compounds **2a** and **2b** this interaction is somewhat hindered when compared to the CFD precursor **1**. In the latter case the *ortho*-hydrogen atoms can freely interact with the solvent molecules which in this particular case of interaction behave as HBA-bases. This interaction can clearly influence the $n \rightarrow \pi^*$ energy of **1**, as this region lies very close to the azobenzene group but also due to π-conjugation between the azobenzene group and the viologen.

Taken together, compound **1**, behaves differently than compounds **2a,b** and constitutes an interesting case where the two solvatochromic functional groups provide significantly different "information" about the polarity effects in their vicinity. In other words, based on SSP-LSERs compound **1** clearly behaves as a solvatochromic compound with dual sensitivity. The overall situation for all three compounds is schematically illustrated in **Figure 9** where the degree of success of each SSP-LSER is marked with colors (**Table 2**).

3.6 Multiprameteric LSERs

An alternative way to compare and quantify the solvatochromism of the two solvatochromic chromophores for **1** and **2a,b** is by employing the multiparametric LSERs. Such LSERs may involve various solvent polarity parameters, each of them

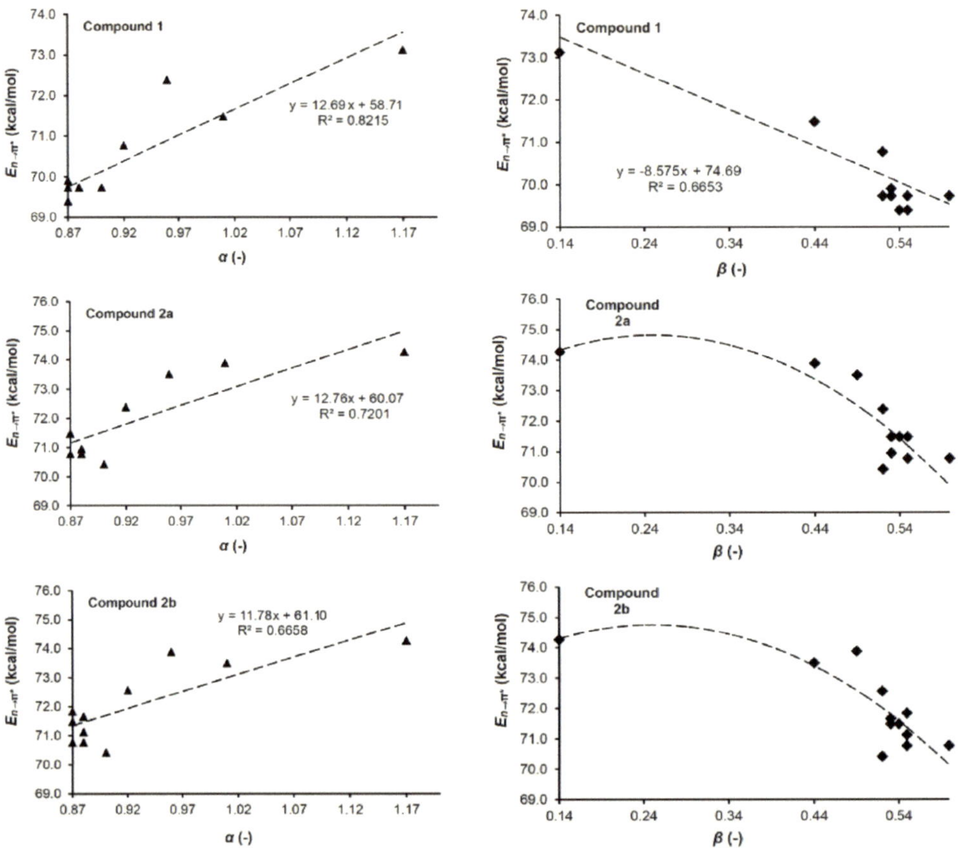

Figure 8.

Left column: Plots of n → π energies determined in aqueous EG mixtures* vs *solvent polarity parameter α for compounds **1**, **2a** and **2b**. Right column: Plots of n → π* energies determined in aqueous EG mixtures* vs *solvent polarity parameter β for compounds **1**, **2a** and **2b** [48].*

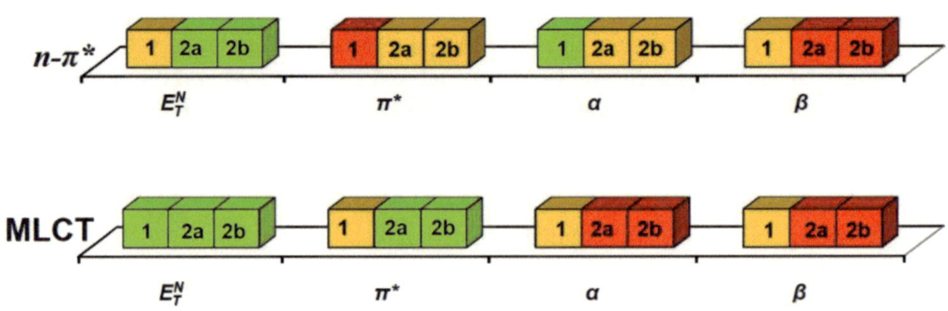

Figure 9.

Illustration depicting the goodness of linear fit between $E_{n\to\pi^*}$ or E_{MLCT} and a SPP (E_T^N, π, α, or β). Colors are based on correlation coefficients of the fits (r^2). Color code: Red: $r^2 < 0.55$; orange $0.55 \le r^2 \le 0.79$; green: $r^2 > 0.79$.*

describing a property of a solvent (or a solvent mixture). A prominent member of the family of such relationships is Kamlet-Abboud-Taft (KAT) equation which can provide information on the solvent polarity effect on various physicochemical properties in terms of specific and non-specific solute-solvent interactions [47]. Deligkiozi et al. employed the triparametric KAT equation (see Eq. 7) and found that for the solvatochromism of the PCF groups (MLCT transitions) in **1** and **2a,b**

	Dimesionless BSM polarity parameters[a]			
x_{EG}	$E_T^N(-)$	$\pi^*(-)$	$\alpha(-)$	$\beta(-)$
0	1.000	1.20	1.17	0.14
0.051	0.971	1.19	1.01	0.44
0.097	0.948	1.18	0.96	0.49
0.139	0.930	1.17	0.92	0.52
0.178	0.916	1.16	0.87	0.53
0.212	0.904	1.15	0.87	0.54
0.245	0.895	1.14	0.87	0.55
0.392	0.861	1.08	0.88	0.53
0.492	0.844	1.04	0.88	0.55
0.659	0.821	0.98	0.87	0.60
0.854	0.806	0.93	0.88	0.55
1	0.790	0.92	0.90	0.52

[a]*The values of the BSM polarity parameters are based on published data [52, 53] and they were determined through polynomial interpolation.*

Table 2.
Solvent polarity parameters employed in this work for the quantification of solvent plarity effects in aqueous EG mixtures.

Compound	E_0†	s†	a†	b†	$\%P_{\pi^*}$	$\%P_\alpha$	$\%P_\beta$	*rse*	r^2
	***n(azo)→π*(azo)* Transitions**								
1	41.70	2.106	23.58	9.370	14.07	51.05	34.88	0.4687	0.8973
2a	36.73	6.425	24.15	11.64	34.70	39.18	26.12	0.2422	0.9762
2b	41.37	6.902	19.94	9.330	40.71	35.11	24.18	0.3461	0.9473
	***dp*(Fe[II])→π*(bpy) Transitions (MLCT)**								
1	5.724	17.50	20.44	4.753	51.27	26.63	22.10	0.6124	0.9630
2a	−11.10	20.42	28.96	13.26	58.56	23.77	17.67	0.3999	0.9854
2b	16.34	22.53	6.521	−2.764	64.66	18.21	17.13	0.4214	0.9845

†*Units: kcal/mol.*
rse: *residual standard error is the square root of the residual sum of squares divided by the residual degrees of freedom (here 8 for all cases).*

Table 3.
Results of the correlation of experimental **n→π*** *and MLCT energies of* **1** *and* **2a-b** *with KAT equation parameters.*

both specific and non-specific interactions are important at various extents depending on the compound (results are summarized in **Table 3**) [19]. Herein, the author reports the corresponding results pertaining to the solvatochromism of the azo group ($n \rightarrow \pi^*$ transitions). By use of Eqs. 8 and 9 (contribution analysis) the relative contribution of each of the parameters π*, α, and β was quantified (detailed results in **Table 3**). Through this analysis it can easily be made clear that for all three compounds the relative importance of the parameters π^*, α, and β, on the energy of the $n \rightarrow \pi^*$ is different when compared to the MLCT transitions. In case of MLCT transitions, parameter π^* appears to contribute the most for all three compounds ($\%P_{\pi^*}$ ranging from 51.27 to 64.66%). Specific solute-solvent interactions described by HBA-basicity (parameter β) and HBD-acidity (parameter α) appear to demonstrate almost equal contribution. In other words, specific and non-specific interactions are almost evenly weighted when it comes to MLCT transitions and that is true for all three compounds. The situation gets way different when interpreting $n \rightarrow \pi^*$ transitions in the same fashion.

The behavior of compound **1** becomes different than those of compound **2a** and **2b**, and the uniformity of the contribution pattern that the MLCT transitions exhibit is lost. While for compound **1** HBD-acidity is the most contributing property (expressed through parameter α) compounds **2a** and **2b** exhibit almost equal contributions of all parameters π^*,α, and β. This is easily explained if the role of CyD (*alpha-* or *beta-*) is taken into account. As analyzed, in these [2]rotaxanes the CyD wheel preferably resides in the region of the azobenzene moiety so as to reduce the interaction with the ionic viologen parts of the axial molecule [54]. This effect hinders solvent molecules from interacting directly and specifically with the azo group (note that the azo group is highly prone to act as HBA-basic group; see **Figure 5B**). In case of **1** i.e. the CFD-precursor of these [2]rotaxanes, the lack of the CyD wheel renders the azo group-solvent interactions highly probable (**Figure 5B**). Therefore, **1** exhibits polarity responsive $n \rightarrow \pi^*$ transitions mainly influenced by parameter α. This constitutes a major differentiation among the studied compound and of course also between the two types of transitions. Obviously, the two transitions discussed herein convey spectrally different polarity information and that holds true for all three compounds but mostly for compound **1**.

3.7 PS effects as sensed by the FC and azo groups

It is well established that when a polar compound is dissolved in a BSM consisted of two solvents of different polarity, the compound/solute gets solvated selectively by one of the two solvents [17, 18]. This effect is obviously associated with preferential solute-solvent interactions developed in the vicinity of the solute molecules. Due to this effect the region around the solute (the so-called cybotactic region) is characterized by a different solvent/cosolvent molar ratio when compared to the bulk part of the solution i.e. the regions away from the cybotactic region. This interesting phenomenon, is attenuated when the two solvents consisting the BSM are similar in terms of structure and polarity [18]. There are numerous published models allowing for the quantification of selective solvation phenomena applicable to various types of solutes and BSMs. These models are generally categorized in thermodynamic and spectroscopy-based models [18]. In the latter case a solvatochromic solute is often employed in order to probe preferential solvation phenomena in BSMs and using spectrally measured shifts as inputs one can obtain various types of information pertaining to preferential solvation as output e.g. the solvent and cosolvent molar ratios in the cybotactic region. Through various spectroscopy-based models thermodynamic properties can also be determined for instance the molar free energy of transfer of the solute from one solvent to its cosolvent [18]. In this work preferential solvation of compounds **1** and **2a,b** in BSMs shall be used as a tool to rationalize the responsiveness of the two types of probing chromophores encompassed in these solvatochromic compounds i.e. the FC and azo groups.

By plotting the experimentally determined MLCT and $n \rightarrow \pi^*$ transition energies of **1** and **2a,b** at various water/EG mole fractions against the mole fraction of water or EG one can easily realize that for both types of transitions a significant deviation from linearity exists (see **Figure 10**). For all compounds the measured energies for either of the transitions MLCT or $n \rightarrow \pi^*$ were lower than the ideal/linear situations

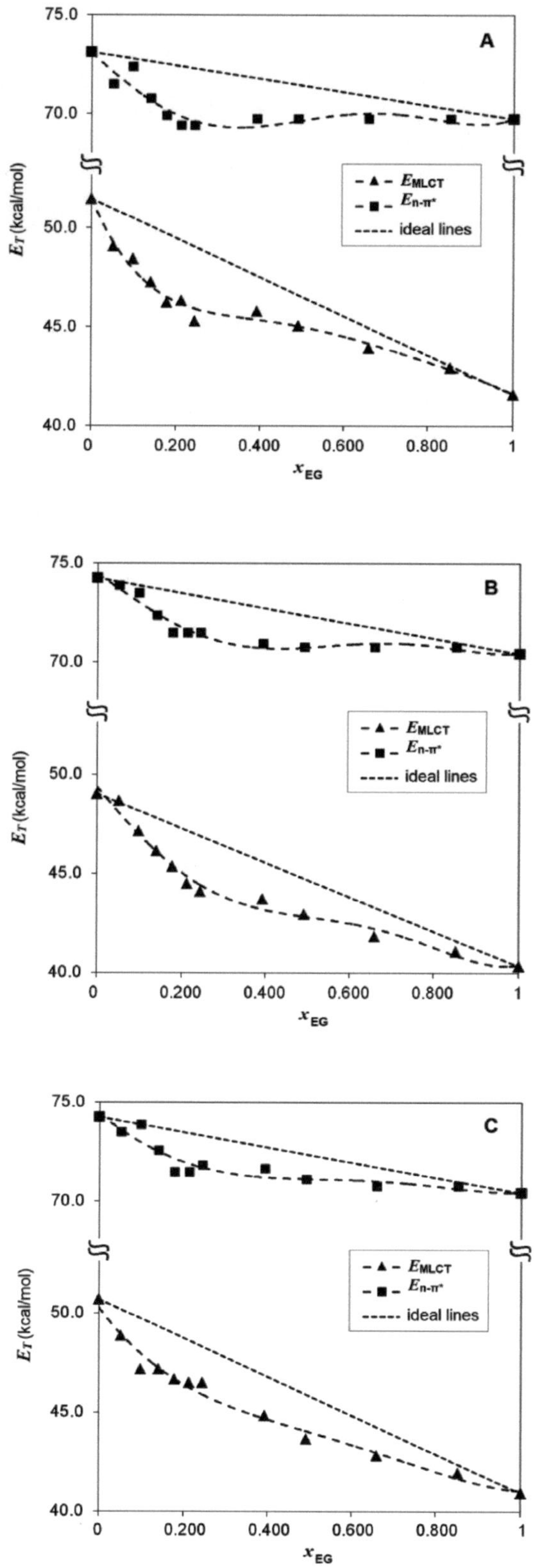

Figure 10.
Plots of experimental transition (MLCT or n→π) energies measured in aqueous EG, against the bulk mole fraction of EG for A) compound 1, B) [2] rotaxane **2a**, and C) [2]rotaxane **2b**.*

(dashed lines in **Figure 10**). Given the fact that both probing chromophores FC and azo, exhibit negative solvatochromism (i.e. increase of transition energy, or hypsochromism, when solvent polarity increases) the plots of **Figure 10** all describe preferential solvation of compounds **1,2a** and **2b** by EG (as will be thoroughly analyzed below). Indeed this effect has been thoroughly discussed in a recent paper by Papadakis et al. pertaining solely to the MLCT transitions. However, the new finding here is that the same effect is beautifully probed also by the azo group at least qualitatively. As can be easily seen, the shape of the non-linear $E_T = f(x_{EG})$ curves of **Figure 10** have very similar shapes for the both types of transitions. It is for instance apparent that the maximization of the linear deviation occurs in all six plots depicted at bulk molar ration of EG $0.15 \leq x_{EG} \leq 0.25$ i.e. in the water-rich bulk solvent composition region. Yet, important differences are revealed when the results are treated quantitatively through the TPMS model by Bagchi and coworkers [32] (details in the Materials and Methods section).

First of all, through the TPMS quantitative treatment the composition of the cybotactic region of solutions of **1** and **2a,b** were determined (see **Table 4**). Compound **1** appears to behave differently than the [2]rotaxanes (**2a,b**). The local EG mole fractions predicted by the same model for the MLCT and azo transition (y_{EG}^{mlct} and $y_{EG}^{n\rightarrow\pi^*}$ respectively) of **1** differ a lot. Interestingly, $y_{EG}^{n\rightarrow\pi^*}$ exhibits a propensity to maximize to the value of **1** (which denotes total solvation by EG) at a very low x_{EG} (x_{EG} = 0.212) whereas y_{EG}^{mlct} follows a much more "legitimate" increase and maximizes only in neat EG. For compounds **2a** and **2b**, y_{EG}^{mlct} and $y_{EG}^{n\rightarrow\pi^*}$ exhibit a very similar increase rate when going from neat water to neat EG (**Table 4**).

This becomes more obvious when plotting the δ_{EG} values obtained through the TPMS model against the bulk EG mole fractions (note: $\delta_{EG} = y_{EG} - x_{EG}$) as predicted through the experimentally observed MLCT and $n\rightarrow\pi^*$ energies. Apparently, compound **1** a different behavior compared to the [2]rotaxanes. The δ_{EG} predicted using the MLCT energies are significantly smaller than those predicted using $n\rightarrow\pi^*$ energies (**Figure 11A**). The difference in δ_{EG} drops when *alpha*-CyD is on (compound **2a, Figure 11B**) and drops even more when *beta*-CyD is threaded around the azobenzene group (compound **2b, Figure 11C**). In fact only δ_{EG} predicted using

x_{EG}	Compound 1		Compound 2a		Compound 2b	
	y_{EG}^{mlct}	$y_{EG}^{n\rightarrow\pi^*}$	y_{EG}^{mlct}	$y_{EG}^{n\rightarrow\pi^*}$	y_{EG}^{mlct}	$y_{EG}^{n\rightarrow\pi^*}$
0	0	0	0	0	0	0
0.051	0.569	0.660	0.510	0.526	0.553	0.555
0.097	0.590	0.561	0.560	0.555	0.610	0.526
0.139	0.635	0.766	0.599	0.662	0.610	0.642
0.178	0.681	0.952	0.634	0.784	0.630	0.784
0.212	0.675	1.000	0.676	0.784	0.637	0.784
0.245	0.728	1.000	0.698	0.784	0.637	0.731
0.392	0.702	1.000	0.719	0.880	0.714	0.757
0.492	0.740	1.000	0.770	0.917	0.782	0.846
0.659	0.807	1.000	0.853	0.917	0.838	0.917
0.854	0.880	1.000	0.922	0.917	0.907	0.917
1	1	1	1	1	1	1

Table 4.
Preferential solvation results obtained through the PS model.

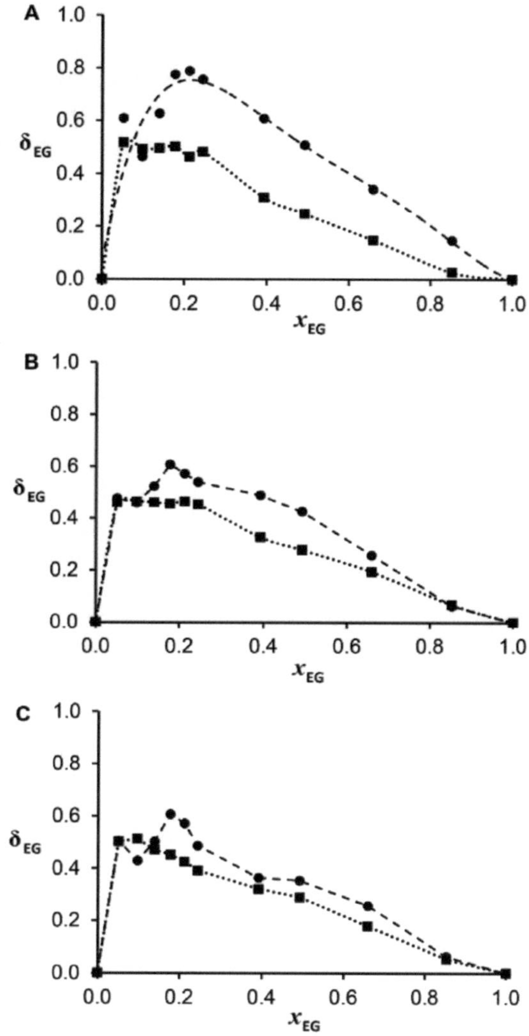

Figure 11.
Plots of TPMS- predicted $\delta_{EG} = y_{EG} - x_{EG}$ versus the bulk EG mole fraction for A) **1**, *B)* **2a** *and C)* **2c**. *Circles and dashed lines correspond to the prediction based on the azo-group whereas squares and dotted lines correspond to the prediction based on the FC group.*

	Compound 1	Compound 2a	Compound 2b
$x_{iso}^{mlct}(EG)$	0.18	0.20	0.30
$x_{iso}^{n\to\pi}$ (EG)	0.055	0.15	0.13
$\Delta \int$	0.18	0.070	0.047
$^{*}\Delta_{mlct}^{n\to\pi}$ (x_{iso})	0.12	0.053	0.17

Table 5.
Isosolvation points and $\Delta \int$ results for compounds **1** *and* **2a,b** *in aqueous EG.*

$n \to \pi^{*}$ energies of compound **1** obtain values as high as 0.8. In other words, the azo group of compound **1** "feels" an excess of EG mole fraction of roughly 0.6 when the bulk EG mole fraction is only 0.2. Such large a preferential effect is not probed by the rest of the compounds. By integrating the differences in δ_{EG} for the three compounds one can easily see a decrease in $\Delta \int$ following the sequence:

$\Delta \int(\mathbf{1}) = 0.19 > \Delta \int(\mathbf{2a}) = 0.071 > \Delta \int(\mathbf{2b}) = 0.046$). Similar conclusions can be drawn when comparing the isosolvation points for the different probes of all compounds (**Table 5**). The results clearly illustrate a different probing aptitude of preferential solvation by the azo group of compound **1**. Of course all results indicate qualitatively the same PS effect i.e. EG is the preferred solvent in the cybotactic region at all measured mole fractions.

4. Conclusions

A general conclusion of the present study is that two distinct functional groups acting as chromophores (specifically the FC and azo groups) can probe solvation effects in different ways however this is true only on a quantitative basis. Qualitatively, both functional groups probed a strong negative solvatochromic effect in all cases of molecules studied. It became apparent though that FC is more sensitive to the dipolarity/polarizability of the medium whereas the azo group is slightly more responsive to polarity changes associated to the Lewis acidity and HBD-acidity of the medium. This holds true for compound **1** and **2a** but not for compound **2b** where the bulkiness of *beta*-CyD hinders any azobenzene-solvent direct interaction. In that case (**2b**) the azo group appears to have very similar sensitivity to solvent polarity to that that the FC group exhibits. On the other hand, both functional groups probed an intense PS effect by EG molecules however for each compound different extents in PS were probed. Compound **1**, again appeared to behave differently on a quantitative basis. The difference in probed extent PS between the azo and the FC group was the largest for **1** and dropped significantly in case of the *alpha*- or *beta*-CyD comprising [2]rotaxanes (compounds **2a**, and **2b**). Overall, compound **1**, exhibits a distinct dual sensing aptitude in terms of solute-solvent specific and non-specific effects as well as PS effects. The [2]rotaxanes have a rather attenuated dual sensing capacity presumably due to the presence of CyD which hinders the direct interaction of the azo-group (and its surrounding regions) with solvent molecules. As a result of the extended π-conjugation and the "shielding" effect of CyD the azo groups of compounds **2a** and **2b** tend to probe the same solvent polarity information as the FC group. It is important to mention that compound **1** clearly exhibits a dual solvatochromic behavior, however in water/EG BSMs the response of the azo group gets saturated to the value corresponding to neat EG very fast as one moves from neat water to neat EG (see **Table 4**). Nonetheless, as analyzed this corresponds to a special PS effect and taken together compound **1**, appears to be a very good polarity sensor candidate for future application mainly pertaining to polar media such as water solutions and mixtures with polar organic solvents.

Acknowledgements

The author would like to thank Dr. D. Matiadis (NCSR Demokritos, Athens, Greece) for fruitful discussions revolving around the solvatochromism of heterocyclic compounds. IKY (Greek State Scholarship Foundation) is gratefully acknowledged for financial support to R.P. during his PhD; a part of this work is connected to the work carried out then.

Notes

The authors declare no competing financial interest.

Author details

Ioanna Deligkiozi[1] and Raffaello Papadakis[1,2]*

1 Laboratory of Organic Chemistry, School of Chemical Engineering, National Technical University of Athens (NTUA), Athens, Greece

2 TdB Labs, Uppsala, Sweden

*Address all correspondence to: rafpapadakis@gmail.com

References

[1] Liu, H.; Xu, X.; Peng, H.; Chang, X.; Fu, X.; Li, Q.; Yin, S.; Blanchard G. J.; Fang, Y. New solvatochromic probes: Performance enhancement via regulation of excited state structures. **Phys. Chem. Chem. Phys., 2016,***18*, 25210-25220. DOI: 10.1039/C6CP04293G.

[2] Ali, R.; Lang, T.; Saleh, S. M.; Meier, R. J. Wolfbeis O. S. Optical sensing scheme for carbon dioxide using a Solvatochromic probe. Anal. Chem., **2011**, *83*, 2846-2851. DOI: 10.1021/ ac200298j.

[3] Landis, R. F.; Yazdani, M.; Creran, B.; Yu, X.; Nandwana, V.; Cooke, G.; Rotello. V. M. Solvatochromic probes for detecting hydrogen-bond-donating solvents. *Chem. Commun.*, **2014**, *50*, 4579-4581. DOI: 10.1039/C4CC00805G.

[4] Florindo, C.; McIntosh, A. J. S.; Welton, T. ; Brancod, L. C.; Marrucho. I. M. A closer look into deep eutectic solvents: exploring intermolecular interactions using solvatochromic probes. *Phys. Chem. Chem. Phys.*, **2018**, *20*, 206-213. DOI: 10.1039/ C7CP06471C.

[5] Liu, H,; Xu, X.; Shi, Z.; Liu, K.; Yu, L.; Fang Y. Solvatochromic probes displaying unprecedented organic liquids discriminating characteristics. Anal. Chem., **2016**, *88*, 10167-10175. DOI: 10.1021/acs.analchem.6b02721.

[6] Li, Z.; Askim, J. R.; Suslick, K. S. The optoelectronic nose: Colorimetric and Fluorometric sensor arrays. Chem. Rev. **2019**, *119*, 231-292. DOI: 10.1021/acs. chemrev.8b00226.

[7] Machado, V. G.; Stock, R. I.; Reichardt, C. Pyridinium N-Phenolate Betaine Dyes. *Chem. Rev.* **2014**, *114*, 10429-10475. DOI: 10.1021/cr5001157.

[8] Reichardt, C. Solvatochromic dyes as solvent polarity indicators. Chem. Rev.**1994**, *94*, 2319-2358. DOI: 10.1021/ cr00032a005.

[9] Cabota, R.; Hunter, C. A. Molecular probes of solvation phenomena. Chem. Soc. Rev., **2012**, *41*, 3485-3492. DOI: 10.1039/C2CS15287H.

[10] Reichardt, C.; Welton. T. Solvents and solvent effects in organic chemistry. 4th Edn. Wiley-VCH, 2011, Weinheim.

[11] Eilmes, A. Kubisiak, P. Explicit solvent modeling of Solvatochromic probes in ionic liquids: Implications of solvation Shell structure. J. Phys. Chem. B, **2015**, *119*, 113185-113197. DOI: 10.1021/acs.jpcb.5b07767.

[12] Khajehpour, M.; Welch, C. M. Kleiner, K. A. Kauffman J. F. Separation of dielectric nonideality from preferential solvation in binary solvent systems: An experimental examination of the relationship between Solvatochromism and local solvent composition around a dipolar solute. J. Phys. Chem. A, **2001**, *105*, 5372-5379. DOI: 10.1021/jp010825a.

[13] Duereh, A.; Sato, Y.; Smith, R. L.; Inomata, H. Spectroscopic analysis of binary mixed-solvent-polyimide precursor systems with the preferential solvation model for determining solute-centric Kamlet–Taft Solvatochromic parameters. J. Phys. Chem. B, **2015**, *119*, 14738-14749. DOI: 10.1021/acs. jpcb.5b07751.

[14] Papadakis, R. Preferential solvation of a highly medium responsive Pentacyanoferrate(II) complex in binary solvent mixtures: Understanding the role of dielectric enrichment and the specificity of solute–solvent interactions. J. Phys. Chem. B, **2016**, *120*, 9422-9433. DOI: 10.1021/acs.jpcb.6b05868

[15] Papadakis, R. Solute-centric versus indicator-centric solvent polarity

parameters in binary solvent mixtures. Determining the contribution of local solvent basicity to the solvatochromism of a pentacyanoferrate(II) dye. J. Mol. Liq. **2017**, *241*, 211–221. DOI: 10.1016/j.molliq.2017.05.147

[16] Papadakis, R. The solvatochromic behavior and degree of ionicity of a synthesized pentacyano (N-substituted-4,40-bipyridinium) ferrate(II) complex in different media. Tuning the solvatochromic intensity in aqueous glucose solutions. Chem. Phys., **2014**, *430*, 29-39. DOI: 10.1016/j.chemphys.2013.12.008.

[17] Ben-Naim, A. Theory of preferential solvation of nonelectrolytes. Cell Biophys. **1988**, *12*, 255-269. DOI: 10.1007/BF02918361.

[18] Marcus, Y. Solvent mixtures: Properties and selective solvation, Marcel Dekker, Inc., 2002, New York.

[19] Deligkiozi, I; Voyiatzis, E.; Tsolomitis, A.; Papadakis, R. Synthesis and characterization of new azobenzene-containing bis pentacyanoferrate(II) stoppered push–pull [2]rotaxanes, with α- and β-cyclodextrin. Towards highly medium responsive dyes. Dyes Pigment., **2015**, *113,* 709-722. DOI: 10.1016/j.dyepig.2014.10.005.

[20] Qu, D-H.; Ji, F-Y.; Wang, Q-C.; Tian H. A double INHIBIT logic gate employing configuration and fluorescence changes. Adv. Mater. **2006**, *18*, 2035-2038. DOI: 10.1002/adma.200600235.

[21] Baroncini, M.; Gao, C.; Carboni, V.; Credi, A.; Previtera, E.; Semeraro, M.; Venturi, M.; Silvi, S. Light control of stoichiometry and motion in pseudorotaxanes comprising a cucurbit[7]uril wheel and an azobenzene-bipyridinium axle. Chem. Eur. J. **2014**, *20*, 10737-10744. DOI: 10.1002/chem.201402821.

[22] Qu, D-H.; Wang, Q-C.; Tian, H. A half adder based on a photochemically driven [2] rotaxane. Angew. Chem. Int. Ed., **2005**, *44*, 5296-5299. DOI: 10.1002/anie.200501215.

[23] Monk, P.M.S. The viologens: Physicochemical properties, synthesis and applications of the salts of 4,40-Bipyridine. John Wiley & Sons ltd; 1998, Chichester.

[24] Crano, J.C.; Guglielmetti, R.J. (Eds). Organic photochromic and thermochromic compounds. Main photochromic families, vol. 1. Kluwer Academic Publishers; 2002. New York. p. 341-67.

[25] Deligkiozi, I.; Tsolomitis, A.; Papadakis, R. Photoconductive properties of a π-conjugated α-cyclodextrin containing [2]rotaxane and its corresponding molecular dumbbell. Phys. Chem. Chem. Phys., **2013**,*15*, 3497-3503. DOI: 10.1039/C3CP43794A

[26] Papadakis, R.; Deligkiozi, I.; Giorgi, M.; Faure, B.; Tsolomitis, A. Supramolecular complexes involving non-symmetric viologen cations and hexacyanoferrate (II) anions. A spectroscopic, crystallographic and computational study. RSC Adv., **2016**, *6*, 575-585. DOI: 10.1039/C5RA16732A.

[27] Papadakis, R; Deligkiozi, I; Tsolomitis A. Synthesis and characterization of a group of new medium responsive non-symmetric viologens. Chromotropism and structural effects. Dyes Pigment., **2012**, *95*, 478-484. DOI: 10.1016/j.dyepig.2012.06.013.

[28] Papadakis, R; Deligkiozi, I; Tsolomitis A. Spectroscopic investigation of the solvatochromic behavior of a new synthesized non symmetric viologen dye: Study of the solvent-solute interactions. Anal. Bioanal. Chem. **2010**, *397*, 2253-2259. DOI: 10.1007/s00216-010-3792-7.

[29] Zhu, Y; Zhou, Y; Wang, X. Photoresponsive behavior of two well-defined azo polymers with different electron-withdrawing groups on pushepull azo chromophores. Dyes Pigment., **2013**, *99*, 209-219. DOI: 10.1016/j.dyepig.2013.05.006.

[30] Papadakis, R.; Tsolomitis, A. Study of the correlations of the MLCT Vis absorption maxima of 4-pentacyanoferrate- 4-arylsubstituted bispyridinium complexes with the Hammett substituent parameters and the solvent polarity parameters ETN and AN. J. Phys. Org. Chem. **2009**, *22*, 515-521. DOI: 10.1002/poc.1514.

[31] Papadakis, R.; Tsolomitis, A. Solvatochromism and preferential solvation of 4-pentacyanoferrate 4-aryl substituted bipyridinium complexes in binary mixtures of hydroxylic and non-hydroxylic solvents. J. Solut. Chem., **2011**, *40*, 1108-1125. DOI: 10.1007/s10953-011-9697-z.

[32] Chatterjee, P.; Bagchi, S. Preferential solvation of a dipolar solute in mixed binary solvent: A study of UV-visible spectroscopy. J. Phys. Chem. **1991**, *95*, 3311-3314. DOI: 10.1021/j100161a064.

[33] Banerjee, D.; Laha, A.K.; Bagchi, S. Preferential solvation in mixed binary solvent. J. Chem. Soc. Faraday Trans., **1995**, *91*, 631-636. DOI: 10.1039/FT9959100631.

[34] Laha, A.K.; Das, P.K.; Bagchi, S. Study of preferential solvation in mixed binary solvent as a function of solvent composition and temperature by UV-vis spectroscopic method. J. Phys. Chem. A, **2002**, *106*, 3230-3234. DOI: 10.1021/jp0121116.

[35] Redlich, O.; Kister, A.T.. Algebraic representation of thermodynamic properties and the classification of solutions. Ind. Eng. Chem.**1948**, *40*, 345-348. DOI: 10.1021/ie50458a036.

[36] Krygowski, T.M.; Fawcett, W. R. Complementary Lewis acid-base description of solvent effects. I. Ion-ion and ion-dipole interactions. *J. Am. Chem. Soc.*, **1975**, *97*, 2143-2148. DOI: 10.1021/ja00841a026.

[37] Papadakis, R.; Deligkiozi, I.; Nowak, E. K. Study of the preferential solvation effects in binary solvent mixtures with the use of intensely solvatochromic azobenzene involving [2] rotaxane solutes. J. Mol. Liq., **2019**, *274*, 715-723. DOI: 10.1016/j.molliq.2018.10.164.

[38] Hofmann, K.; Brumm, S.; Mende, C.; Nagel, K.; Seifert, A.; Roth, I.; Schaarschmidt, D.; Lang, H.; Spange, S. Solvatochromism and acidochromism of azobenzene-functionalized poly(vinyl amines). New J. Chem., **2012**, *36*, 1655–1664. DOI: 10.1039/c2nj40313g.

[39] Sıdır, Y. G.; Sıdır, I; Taşal¸ E., E. Ermiş. Studies on the electronic absorption spectra of some monoazo derivatives. *Spectrochim.* Acta A **2011**, *78*, 640–647. DOI: 10.1016/j.saa.2010.11.0.

[40] Gasbarri, C.; Angelini, G. Polarizability over dipolarity for the spectroscopic behavior of azobenzenes in room-temperature ionic liquids and organic solvents. J. Mol. Liq., **2017**, *229*, 185–188. DOI: 10.1016/j.molliq.2016.12.033.

[41] Qian, H-F.; Tao, T.; Feng, Y-N.; Wang, Y-G.; Huang, W. Crystal structures, solvatochromisms and DFT computations of three disperse azo dyes having the same azobenzene skeleton. J. Mol. Struct. **2016**, *1123*, 305-310. DOI: 10.1016/j.molstruc.2016.06.04.

[42] Christoforou, D. Electronic effects of the azo group. PhD Thesis, University of Canterbury, New Zealand, 1981.

[43] Mulski, M. J.; Connors, K.A. Solvent effects on chemical processes. 9. Energetic contributions to the

complexation of 4-nitroaniline with α-cyclodextrin in water and in binary aqueous-organic solvents. **1995**, *4*, 271-278. DOI: 10.1080/10610279508028936.

[44] Hickey, J.P.; Passlno-Reader D.R. Linear solvation energy relationships: "Rules of thumb" for estimation of variable values. Environ. Sci. Technol., **1991,** *25*, 1753-1760. DOI: 10.1021/es00022a012.

[45] Endo, S; Goss, K-U. Applications of Polyparameter linear free energy relationships in environmental chemistry. Environ. Sci. Technol., **2014**, *48*, 12477–12491. DOI: 10.1021/es503369t.

[46] Williams, A. Free Energy Relationships in Organic and Bio-Organic Chemistry. Royal Society of Chemistry, Cambridge UK, 2003.

[47] Kamlet, M.J.; Abboud, J.L.M.; Abraham, M.H.; Taft, R.W. Linear solvation energy relationships. 23. A comprehensive collection of the solvatochromic parameters, .Pi.*, . Alpha., and.Beta., and some methods for simplifying the generalized solvatochromic equation, *J. Organomet. Chem.* **1983**, *48,* 2877–2887. DOI: doi. org/10.1021/jo00165a018.

[48] The non-linear fit(s) in this figure correspond to polynomial fitting(s) and convey no physical meaning. They are simply used in order to visualize the sizable deviations from linearity.

[49] Ratts, K. W.; Howe, R. K.; Phillips, W. G. Formation of pyridinium ylides and condensation with aldehydes. J. Amer. Chem.Soc.**1969**, *91*, 6115-6121. DOI: 10.1021/ja01050a032.

[50] Zoltewicz, J. A.; Smith, C. L.; Kauffman, G. M. Buffer catalysis and hydrogen-deuterium exchange of heteroaromatic carbon acids. Heterocycl. Chem. **1971**, *8*, 337-338. DOI: 10.1002/jhet.5570080236.

[51] Elvidge, J.A.; Jones, J.R.; O'Brien, C.; Evans, E.A.; Sheppard, H.C. Base-Catalyzed Hydrogen Exchange. *Adv. Heterocycl. Chem.* **1974**, *16*, 1-31. DOI: 10.1016/S0065-2725(08)60458-4.

[52] Marcus, Y. The use of chemical probes for the characterization of solvent mixtures. Part 2. Aqueous mixtures. *J. Chem. Soc.* Perkin Trans. **1994**, *2*, 1751–1758. DOI: 10.1039/P2994000175

[53] Sindreu, R. J.; Moyá, M. L.; Sánchez Burgos, F.; González, A. G. Kamlet-Taft solvatochromic parameters of aqueous binary mixtures oftert-butyl alcohol and ethyleneglycol. J. Solut. Chem. **1996**, *25*, 289-293. DOI: 10.1007/BF00972526.

[54] Deligkiozi, I.; Papadakis, R.; Tsolomitis, A. Synthesis, characterisation and photoswitchability of a new [2]rotaxane of α-cyclodextrin with a diazobenzene containing π-conjugated molecular dumbbell. Supramol. Chem. **2012**, *24*, 333-343. DOI: 10.1080/10610278.2012.660529.

Chapter 3

Dyes as Labels in Biosensing

Hu Li, Yuanyuan Han, Haiyan Zhao, Hassan Jafri and Bo Tian

Abstract

Investigation and evaluation of dyes play a vital role in the process of introduction novel labels and their corresponding sensing methods, which signify opportunities for the development of biosensors. This chapter introduces applications of various dyes as labels in biosensing. Bio-recognition molecules with dyes transduce biological information into measurable optical, electrochemical, magnetic or other kinds of signals for quantification. The dyes used in this field were summarized and reviewed according to their signal types, namely colorimetric, fluorescent and electrochemical. Some dyes can transduce signals between multiple physical signals. For some most important dyes, detailed descriptions were given focused on their unique properties, labeling methods and sensing mechanisms.

Keywords: biosensors, labels, micro- and nano-particles, optical dyes, charge-transfer complexes

1. Introduction

Applications of, e.g., clinical diagnosis, drug development, environmental research, security and defense, require self-contained rapid analytical platforms to get rid of tedious operation processes and long turn-around times. In 1960s, applications of enzymes explored a new way in analytical chemistry to obtain specific, sensitive and ease-of-use assays. At the same time, ion selective electrodes have been developed for rapid non-reagental analysis of inorganic ions. In 1962, the concept of enzyme transducer was proposed and sooner the device was developed [1]. Following this idea of enzyme electrode, the first enzyme electrode-based glucose meter was commercialized in 1975. After that, self-contained analytical platforms based on different principles were developed including thermistor, [2] optical fiber, [3] piezoelectric crystal detector, [4] surface plasmon resonance, [5] etc. Today, such analytical platforms are regarded as biosensors. A biosensor is a self-contained integrated device that is capable of providing specific quantitative or semi-quantitative analytical information using a biological recognition element (biochemical receptor) which is in direct spatial contact with a transducer element [6]. Today, biosensors have been applied to a wide variety of analytical problems, e.g., medicine, environmental research, food control, and process industry [7].

Biosensors can be divided into two groups based on involving a labeling process during the detection or not. A label is a foreign molecule that is chemically or temporarily attached to the target (i.e., the molecule of interest) through a labeling process to detect molecular presence or activity. In the above-mentioned early

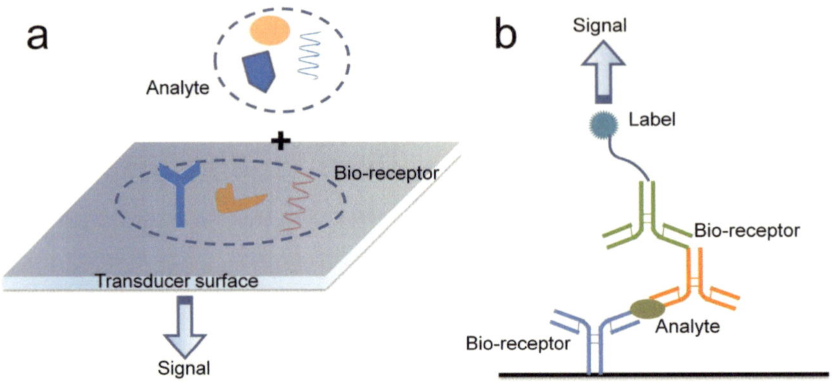

Figure 1.
Scheme of typical (a) label-free and (b) label-based biosensors.

period of the biosensor history, typical biosensing processes were usually realized by measuring the transduced mechanical, electrical, or optical signals without any labels, as illustrated in **Figure 1a**. Such label-free biosensors can provide direct information without complicated sample preparation steps. In contrast, label-based biosensors utilizing additional operation processes for higher signal-to-noise ratios and for a broader range of sensing/transducer systems. Conventional biosensing labels are optical molecules or radioactive elements borrowed from bio-analytical systems such as gel electrophoresis and enzyme linked immunosorbent assays. A typical strategy of sandwich assay for antibody-based detection is shown in **Figure 1b**.

Due to the fast development of nanotechnology and material science, nanomaterials are widely adopted as biosensing labels, many of which are dyes in senses of optics or electrochemistry. The special size of these nanomaterial dyes provides unique properties that greatly improve the performances of relevant biosensors. In this chapter, we focus on dyes used as biosensing labels and discuss their properties, applications and how they improve the biosensing properties.

2. Dyes in colorimetric biosensors

Colorimetry technique is a practical and direct analytical method to determine the concentration of colored analyte depending on the color change in solution. The colorimetric biosensing strategy based on this principle has become one of the most popular and important strategies due to its simplicity, visualization, low cost and non-destruction. Based on the strategies of signal generation, colorimetric assays can mainly be divided into two groups, i.e., assays based on enzymes for chromogenic reactions, and assays based on colored labels. Herein, we focus on the later group in which dyes play a crucial role.

2.1 Gold nanoparticles

As a successful example to employ nanomaterial and nanotechnology to solve the biological problems, colloidal gold has been introduced in the biosensing fields more than 20 years with lots of amazing works being reported. Colorimetric biosensors based on gold nanoparticles (GNPs) have greatly developed in both scientific study and commercial applications.

2.1.1 Sensing mechanism

Surface plasmon resonance. Various GNP-based colorimetric biosensors were built depending on the same principle, i.e., surface plasmon resonance (SPR). SPR is a prominent spectroscopic feature that results in an intense and sharp absorption band in the visible range of noble metal nanoparticles that have an adequate density of free electrons. Localized SPR is an effect that the electron cloud of the nanoparticle sense and start to oscillate at the same frequency as the electromagnetic field of the incident light (**Figure 2a**). During this process the incident light is scattered and converted into heat, both leading to the intensity attenuation of the incident light. Meanwhile, localized SPR produce an electric field on the nanomaterial surface, which can be utilized for labeling several kinds of biosensors. The SPR-induced color is determined by several factors including the size, shape, modified ligands and aggregation state. In colorimetric biosensors, the aggregation state of GNPs is interested. As shown in **Figure 2b**, in bioanalytical assays, the combination of bio-receptors (or targets) labeled with GNPs can induce the isolated GNP assembly as well as the interparticle coupling of the surface plasmon, resulting in the color of solution transferring from red to blue.

Surface-enhanced Raman spectroscopy. Based on the SPR effect, GNPs have also been used in other types of biosensors, e.g., surface-enhanced Raman spectroscopy (SERS) based sensing. SERS as a versatile finger-print vibrational technology has been widely utilized in analytical chemistry, [8] electrochemistry, [9] and media diagnostics. [10] However, the mechanism of SERS is still

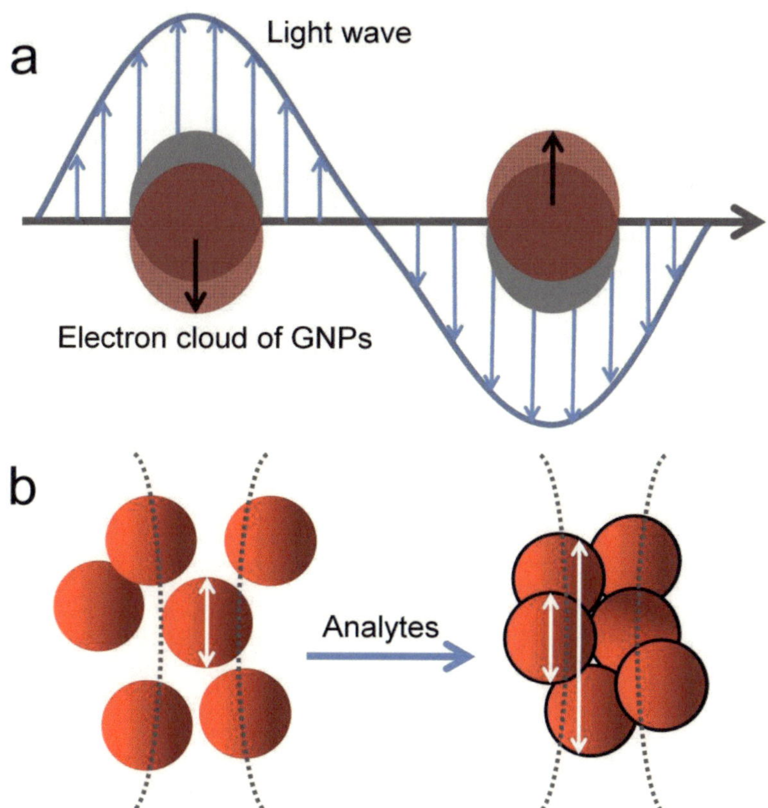

Figure 2.
(a) Localized SPR band formation. Red and gray circles represent negative and positive electron clouds, respectively. (b) GNP agglomeration leads to the shift of SPR absorption band.

under debating [11]. Regarding to the mechanism of SERS, there are two broadly accepted opinions: electromagnetic (EM) theory and chemical enhancement (CE) mechanism [12]. The EM enhancement considers the molecules as point dipoles, which react with the local enhanced electric field on or near the surface [13]. The CE mechanism attributes the SERS intensity to molecular resonance through the interaction with the metal surface, thereby changing the molecular polarizability, resulting in enhancement such as resonance Raman scattering [14]. SERS does not occur on any metal surfaces, among which GNPs are currently the most widely used SERS substrate. SERS-based biosensors involving GNPs have been widely used in the sensitive and selective detection of antigens, [15] aptamers, [16] tumor biomarkers, [17] as well as *Staphylococcus aureus* [18].

Lateral flow system. The most well known biosensing application of GNPs must be the pregnancy test sticks, which belong to the most commercialized type of biosensor, i.e., the lateral flow (LF) system. LF biosensors are paper-based devices permit low-cost and rapid diagnostics with moderate robustness, specificity and sensitivity. The ease-of-synthesize, stability, biocompatibility and tunable size make GNPs suitable label for LF systems. Moreover, the SPR effect endows GNPs an intense red color that can be either detected qualitatively by naked eye or measured quantitatively using spectrometers for lower detection limits. Because of these properties, GNPs are the most widely used optical label in LF systems [19]. A standard and conventional LF strip consists of four main sections made of membranes, papers or glass fibers, including a sample pad for sample loading, a conjugate pad impregnated with bioreceptor-modified labels (usually GNPs), a detection pad/membrane where test line (to show whether the target is exist in the sample) and control line (to show whether the LF assay works well) are printed, and an absorption pad at the end of the strip, as illustrated in **Figure 3**. Except for GNPs, other colored materials are also used in LF biosensors, including carbon dots and latex particles.

2.1.2 Labeling type

Three principle methods for modifying GNPs are briefly illustrated in **Figure 4**: (1) A simple and common way to immobilize the biomolecules on the GNPs is labeling through the sulfur-gold interactions; [20] (2) Ligands can be capped on the GNPs in the process of particle growth or though ligands replace after the synthesis; (3) Ligands can be absorbed on the GNPs surface directly via non-covalent interactions.

In 1997, the first GNP-based colorimetric biosensor was reported [21]. In this work, the hybridization of DNA probes capped on GNP surface and target DNA

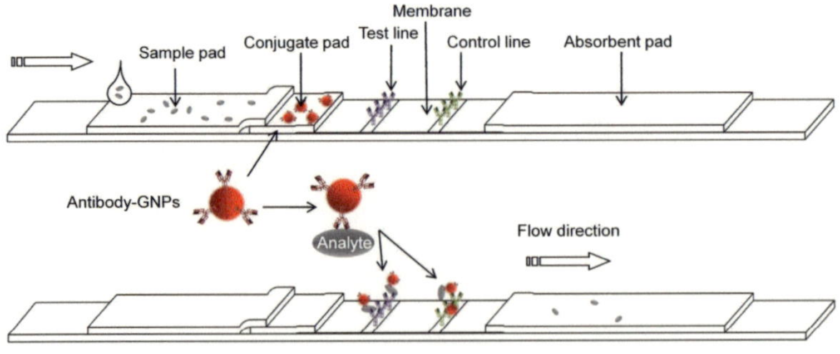

Figure 3.
Schematic of a typical lateral-flow immunological detection system.

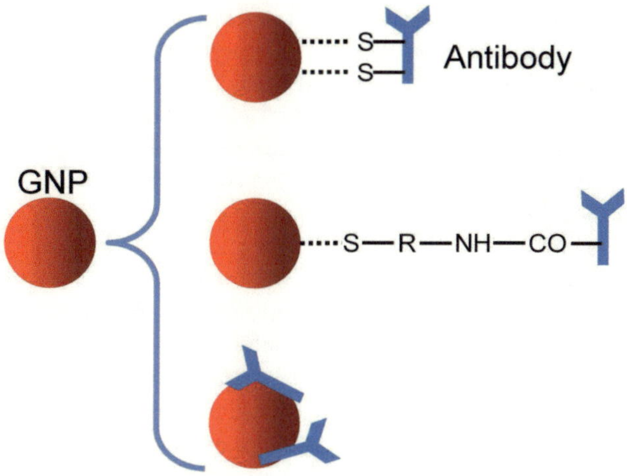

Figure 4.
Illustration of different GNP conjugation methods.

resulted in the formation of GNP agglomeration, accompanying a visually red-to-purple color change. This is the most typical biosensor that uses biochemical reactions to induce the cross-linked aggregation of GNPs. Since then, a new generation of medical diagnostic technology based on the nanomaterials has begun. Oligonucleotides, peptides, antibodies and aptamers have been labeled with GNPs for the colorimetric detection of different targets based on the similar principle. Such strategy designs usually achieve naomolar detection limits, which is limited by the signal-to-noise ratio related to the intrinsic properties of labels and sensors [22]. To amplify the signals and optimize the performance of biosensor, the biochemical and molecular amplification methods are introduced in the biosensing process. Via duplex-specific nuclease-assisted amplification method, a colorimetric method was developed for microRNA dection based on GNPs aggregation [23]. DNAzyme-assisted target recycling was utilized, combined with surface plasmons of GNPs coupling in the colorimetric biosensor, obtaining a fast and simple detection of genetic targets with 50 pM sensitivity [24]. Using imaging-based analysis instead of spectrographic analysis has higher signal-to-noise ratio and thus potentially lower detection limit. A dark-field microscope based methodology was applied for the sensing of GNP aggregation, obtaining a detection limit of 43 aM of DNA, which was 5–9 orders of magnitude lower than conventional colorimetric sensor based strategies [25].

2.2 Carbon nanoparticles

Carbon nanoparticles, also named colloidal carbon, can be visually detected in a qualitative or semi-quantitative manner and thus being used as colored labels. Compared to GNPs, carbon nanoparticles have several excellent properties, e.g., high stability, nontoxicity, ease-of-preparation, and ease-of-modification [26]. The dark black color of carbon dots endows a high signal-to-noise ratio, allowing sensitivities below the low picomolar range even by visual inspection [27].

2.3 Latex particles

Colored latex particles are also often used as labels in the colorimetric biosensor. Latex particles are natural or synthetic polymer nano- and micro-particles that suspend stably in water, and the polystyrene particles are used mostly.

There are three ways to prepare colored latex particles by dying latex particles with different types of dyes molecules: (1) co-polymerization of polymer monomer with dyes; (2) cross-linking the dyes on particles surface by covalent bonds; (3) physical embedding or absorption dyes in particles. After the dying, usually dyes on the surface are removed in order to functionalize the active groups (sulfhydryl groups, amino groups and carboxyl groups) on the latex particles for further labeling the biomolecules [28].

Benefit its wide variety of sources, low cost and easy to be functionalized, the latex particles are applied as probes in immunochromatographic analysis quite early [29]. The good properties enable them are still used now. A lateral flow immunoassay was developed by covalent functionalizing the antigens on colored latex particles for the visual diagnosis of canine visceral leishmaniasis [30]. A latex particles-GNPs composites labeled with antibodies were synthesized as probed for the immunochromatographic test. The nanocomposites amplified the binding capacity of GNPs with target antigens and improved the sensitivity 2 orders of magnitude compared with GNPs-antibodies probes [31].

3. Dyes in fluorescent biosensors

In the field of biotechnology, diagnosis and drug discovery, fluorescent assay is by far the most popular methodology because of not only its sensitivity and versatility but also the high commercialization of fluorescent labels [32]. In addition to the new fluorescent nanomaterials (e.g., upconversion fluorescent materials and aggregation-induced emission (AIE) materials that are described further below), new spectroscopic sensors have also been developed based on rising technologies such as fiber optics, LEDs and fast imaging devices, all of which contributed to the fast development as well as high interdisciplinarity of fluorescent biosensors.

3.1 Organic dyes

Organic fluorescent dyes are a class of organic molecules that contain a fluorescent core skeleton with a large conjugate system and some auxochrome or active group (such as carboxyl, amino, amide, etc). The fluorescent core skeleton enables them absorb a certain excitation light and emitting it as fluorescence. The auxochrome or active group is capable of altering wavelengths and enhance fluorescence or labeling them to bio-receptors for recognizing various biomolecules in biosensing [33]. Briefly, the fluorescent dyes labeled bio-receptors, also called fluorescent probes, can recognize various biomolecules and then convert the recognition events into fluorescent signal output to achieve biosensing or imaging.

Currently, there are many kinds of organic fluorescent dyes, most of which can be used to label bio-receptors for biosensing and imaging. Here, some major organic fluorescent dyes labels are introduced, including fluorescein derivatives, rhodamine derivatives, cyanine derivatives and other commonly used organic fluorescent dyes.

3.1.1 Sensing mechanism

The signal conversion mechanism in sensing process are various, such as electron transfer quenching or fluorescence recovery, fluorescence resonance energy transfer (FRET), or monomer-excimer emission conversion with pyrene fluorophores. The following will introduce the major signal conversion mechanism involved in the fluorescence biosensing process.

Nucleic acid intercalating dyes. Nucleic acid intercalating dyes [34–36] is a special kind of organic fluorescent dyes that have no fluoresce or the fluorescence is weak in solution, which may be caused by the quenching of solvent. However, when they are embedded in specific DNA structures, the fluorescence intensity will increase significantly, due to the protection of the hydrophobic groups of the oligonucleotide (**Figure 5**). Commercially available nucleic acid intercalating dyes for labels in biosensing mainly include ethidium bromide (EB), thioflavin T (ThT), N-methylporphyrin dipropionic acid IX (NMM) and triphenylmethane dyes. The most classical biosensing application is EB staining-mediated gel electrophoresis for nucleic acids detection [37]. Nucleic acid fragments can be separated in gel under the action of electric field, then EB contains a tricyclic planar group can insert between nucleic acid stacking bases, resulting in increased fluorescence intensity of EB for detection.

An emerging biosensing strategy is designed based on the G-quadruplex (G4) DNA structure and corresponding intercalating dyes such as ThT, NMM, etc. (**Figure 6**) [38]. G4 DNA structure is formed from DNA guanine-rich sequences, which has been confirmed to be stably present in human live cell [39]. Therefore, endogenous G4 DNA can be easily detected by using intercalating dyes targeting G4 DNA. Additionally, G4 DNA structures can be formed by the amplicons of any kinds of DNA amplification methods that produce single-stranded DNA, making G4 structure a convenient cascade amplification tool (a molecular amplification followed by a signal amplification) that can be applied in homogeneous and isothermal bioanalytical assays. Moreover, the formation or consume of the G4 structure after binding to the target molecules will change the interaction between G4 and intercalating dyes, resulting in increased or decreased fluorescence intensity for detection.

Fluorescence resonance energy transfer. Measuring the presence of labels always means employing tedious operation steps for separation and washing. Today, homogeneous reaction processes are highly preferred due to its potential for point-of-care applications. FRET assays are frequently used in biosensors due to achieve homogeneous reaction processes with high sensitivity.

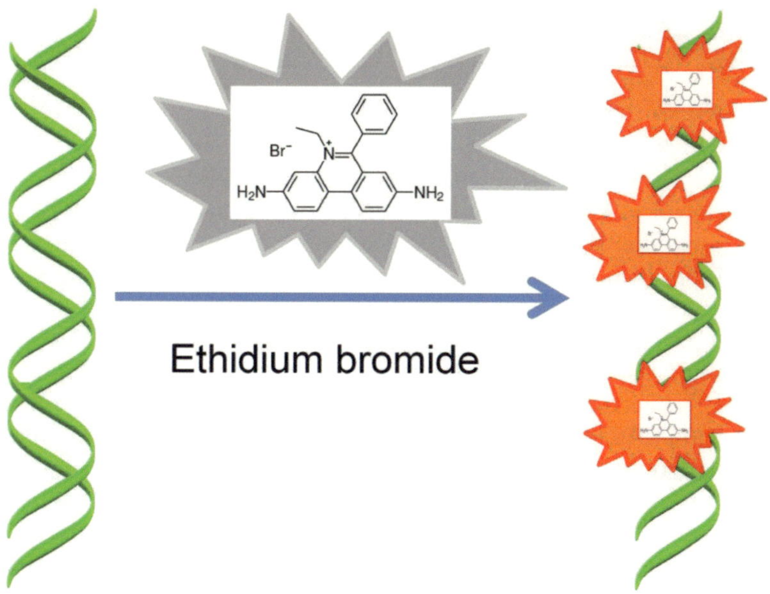

Figure 5.
EB staining for nucleic acids detection.

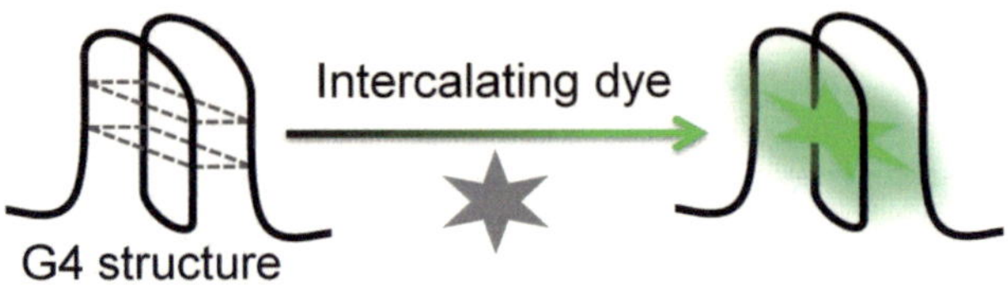

Figure 6.
Fluorescence enhancement based on the intercalating dyes and G4 structure.

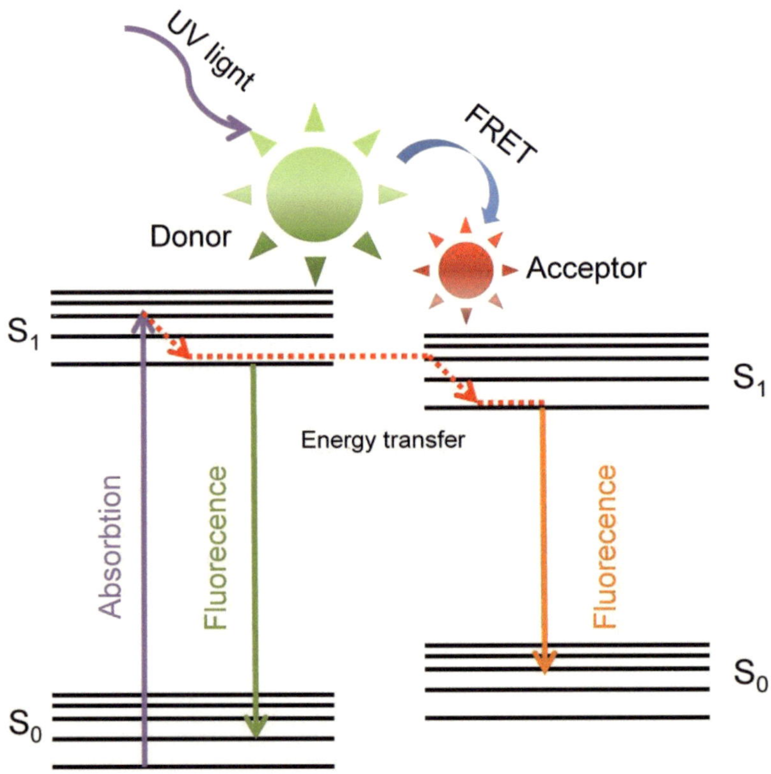

Figure 7.
Schematic illustration of fluorescence resonance energy transfer.

A FRET technique includes an energy transfer between two fluorophores, i.e., from high energy donor to a lower energy acceptor (**Figure 7**). The FRET occurs when the donor and acceptor are close to each other, approximately between 1–10 nm, and this distance meets the dimensions of biological molecules. Since the FRET is sensitive to the relative distance between donors and acceptors, when biological acceptor labeled with donors (or acceptors) comes close to the biological target labeled with acceptors (or donors), FRET signals can be detected.

Rather than labels the donor and the acceptor on different biomolecules, a molecular beacon (MB) utilized a donor linked with an acceptor through a bioresponsive probe, further simply the design of FRET biosensors [40]. Typically, a MB is a single-stranded oligonucleotide probe labeled with a fluorophore at its one end and a quencher at the other. Due to the length and/or the secondary structure of the oligonucleotide, the probe holds the fluorophore and quencher close to each other, thus inducing a quenching. Upon the hybridization between the probe and the target, the distance between the fluorophore and the quencher changes, restoring the fluorescence [41] (**Figure 8**).

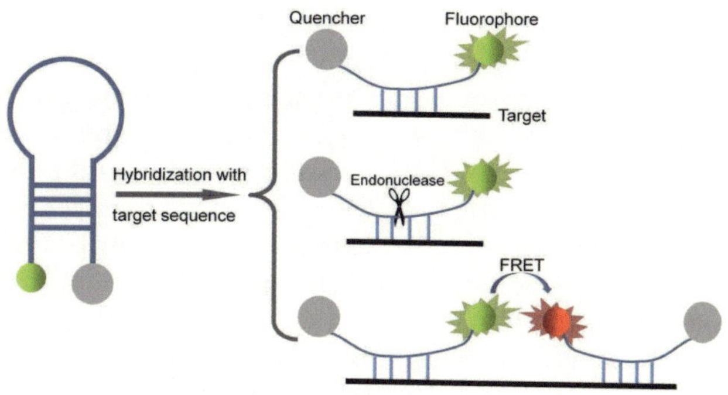

Figure 8.
Illustration of a molecular beacon and examples of its applications.

Monomer-Excimer. Generally, when the distance and position between fluorescent dyes with the same or different structures are appropriate, the excited state fluorescent dye and the other ground state fluorescent dye would form an excimer. Therefore, the fluorescence emission intensity of the original monomer weakens or disappears, and the fluorescence emission of the formed excimer appears [42, 43]. For monomer-excimer based biosensing process, the reaction between the fluorescent dyes-labeled probe and the target biomolecules will trigger or hinder the monomer-excimer process, causing the variation of fluorescence emission spectra.

3.1.2 Labeling type

The labeling type between organic fluorescent dyes and bio-receptors can generally be divided into two types: covalent and non-covalent (such as intercalation, groove binding or electrostatic interaction). Different labeling method has different fluorescent sensing mechanisms.

Covalently binding labeled fluorescent probes including single-labeled fluorescent probe and dual-labeled fluorescent probe. Single-labeled fluorescent probes are obtained by covalently binding single fluorescent dye molecules to bio-receptors. The single-labeled probe sensing mechanism may be summarized as follows: when the bio-receptor of the fluorescent probe recognizes the target molecules, the fluorescence properties of dyes would be changed, such as changes in fluorescence intensity and fluorescence anisotropy, thus converting the recognition process into a measurable fluorescence signal [44–46]. Dual-labeled fluorescent probes are obtained by covalently binding dual fluorescent dye molecules (or a dye and a quencher) to bio-receptors. The dual-labeled probes are all distance dependent, the rearrangement of the probe structure after binding to the target molecules will change the distance between the two labels, resulting in changes in the fluorescence properties of the system.

Non-covalently binding fluorescent probes mainly refer to nucleic acid fluorescent probes that obtained by the binding of intercalating dyes and DNA. As mentioned above, when the nucleic acid intercalating dye binds to a specific DNA structure, the fluorescent signal changes. Based on this principle, a series of biosensors have been developed. Compared with covalently binding fluorescent probes, the non-covalently binding fluorescent probes will not affect the binding affinity of the probe to the target, and also have the advantages of easy operation and low cost [47].

3.1.3 Fluorescein derivatives

Fluorescein and its derivatives are one class of xanthene dyes. Fluorescein was first produced by Von Bayer in 1871, which has a good rigid coplanar structure and can produce strong green fluorescence under the action of laser. Due to its easy synthesis, low cost, low biological and cytotoxicity, high molar absorption coefficient, and high fluorescence quantum yield, fluorescein can be widely used in biological imaging and analysis [48]. However, fluorescein also has some defects, such as high pH sensitivity [49], small stokes [50] and poor light stability [51]. In order to improve the fluorescence performance of fluorescein, many important fluorescein derivatives have been developed by introducing functional group modification to fluorescein [52]. Additionally, fluorescein derivatives contain some active groups, which can bind with bio-receptors to obtain fluorescent dyes labeled probes with high selectivity, good stability and high sensitivity for biosensing [53]. Currently, commercially available fluorescein derivatives dye mainly includes 6-carboxy-fluorescein (FAM), 5-tetrachloro-fluorescein (TET), 5-hexachloro- fluorescein (HEX).

3.1.4 Rhodamine derivatives

It was discovered in the late 1980s that rhodamine and its derivatives are important fluorescent dyes and also belong to xanthene dyes. The molecular structure of rhodamine dyes is very stable, coplanar, and can produce strong red fluorescence under the excitation. They also can bind with bio-receptors to obtain fluorescent dyes labeled probes with high selectivity, good stability and high sensitivity for biosensing. Compared with fluorescein derivatives, rhodamine derivatives have stronger photostability, higher fluorescence quantum yield and lower pH sensitivity. Commercially available rhodamine derivatives dyes for labels in biosensing mainly include 6-carboxyl-x-rhodamine (ROX), tetramethyl-6-carboxyrhodamine (TAMRA) and Texas red.

It was discovered in the late 1880s that rhodamine and its derivatives are important fluorescent dyes and also belong to xanthene dyes [54]. The molecular structure of rhodamine dyes is very stable, coplanar, and can produce strong red fluorescence under the excitation [55]. They also can bind with bio-receptors to obtain fluorescent dyes labeled probes with high selectivity, good stability and high sensitivity for biosensing. Compared with fluorescein derivatives, rhodamine derivatives have stronger photostability, higher fluorescence quantum yield and lower pH sensitivity [56]. Commercially available rhodamine derivatives dyes for labels in biosensing mainly includes 6-carboxyl-x-rhodamine (ROX), tetramethyl-6-carboxyrhodamine (TAMRA) and Texas red.

3.1.5 Cyanine derivatives

Cyanine dyes was first discovered by Williams in 1856, subsequently, Vogel discovered that these dyes have very good photosensitivity, which promote the development of cyanine dyes [57]. Cyanine dyes and derivatives have excellent spectral characteristics, such as high molar extinction coefficient, high fluorescence quantum yield, and long fluorescence emission wavelength. More importantly, the maximum emission and absorption of these dyes are located in the near-infrared region. In this region, the self-absorption and background fluorescence of biomolecules are relatively small [58]. Thus, cyanine dye derivatives have become the most commonly used fluorescent signal groups in protein, nucleic acid and other biological analysis [59]. Commercially available cyanine derivatives dyes for labels in biosensing mainly refers to N-carboxypentyl-5-sulfonate-3H-indocyanine dyes, including Cy3, Cy5, Cy5.5 and Cy7.

3.2 Quantum dots

Quantum dots (QDs) are spherical or quasiballistic semiconductor nanoparticles that bind excitons in the three dimensions, with a diameter no larger than twice the Bohr radius of the excitons of their corresponding semiconductor material, thus confining the motion of electrons, holes, or excitons in three dimensions. Due to the quantum confinement effects, the quantum dots display unique optical and electronic properties compared to the bulk materials.

QDs were firstly synthesized in glass matrix in 1970s and with their fluorescent properties reported [60]. Later some groups studied the novel properties of quantum dots and tried to investigate influences of quantum effects on the optical properties of QDs [61]. In 1980s, CdS nanospheres were synthesized in colloidal solution and the basic theory of QDs were studied [62]. In 1993, the high quality colloidal QDs were prepared first time with uniform size in the solution [63], which provided favorable materials for both theoretical study and practical applications. Since then, various types of QDs with different compositions and properties have been synthesized by the solution growth method.

According to the chemical compositions, QDs can be mainly divided to two groups. Single component QDs, such as metal chalcogenides, [64] attracted much attentions at the early stage of the QDs development. Due to the uniformity, optical and electronic properties of such QDs can be tuned by simply controlling their sizes. Multiple component QDs are either core-shell structural or alloyed. Core-shell structural QDs have a core with one component embedded in another material as a shell, such as CdSe/ZnS [65]. Usually, to reduce the nonradioactive recombination of electron–hole pairs, the material used as the shell has a larger band gap than that in the core, thus improving the fluorescence quantum yield. Coating the same core with different shells adjust the properties of the QDs. Alloyed QDs that have homogeneous or heterogeneous alloyed internal composition, for example, CdSxSe1-x/ZnS [66]. This type of QDs allows tuning the properties by changing the proportion of the component without changing the size. Interestingly, alloyed QDs not only exhibit the original properties of each component, but also display newly additional and adjustable properties because of the merge of the different composites. Now despite classical nanocrystals, there are various new species of QDs that have been prepared, such as perovskite QDs [67] and graphene QDs [68].

Generally, the size of QDs, i.e., in the range of 2–10 nm, endows these nanoparticles high surface-to-volume ratios. The large surface provides rich sites for further functionalization and immobilization of molecules, including nucleic acids and proteins. [69] Importantly, after functionalized with hydrophilic ligands, QDs are soluble and stable in aqueous solution, which is the common environment for biological reactions.

3.2.1 Sensing mechanism

The most obvious and widely utilized properties of QDs are the optical properties. Compared with organic dyes, QDs display higher fluorescence quantum yield and extinction coefficient, broader excitation spectra, longer lifetimes and tunable fluorescent emissions [70]. As the size of QDs decreases, the band gap between valence band (VB) and conduction band (CB) increases, which means more energy needed for electrons excitations (from ground state in VB to CB) as well as more energy released from the electrons de-excitation (from CB to ground state inVB), leading to the fluorescent emission shift to the high frequency range. In addition, the fluorescence wavelength can also be tuned flexibly by adjusting the structure and chemical compositions of QDs as mentioned above. These stand out properties make QDs appealing for bio-medical applications including imaging and biosensing [71]. In biosensing systems, QDs, used with or without nano-sized quenchers, are transducers and functionalized with bio-recognition molecules (bio-receptors). Because of the tunable size and broad spectral width, QDs can play as either energy donors or acceptors in the FRET biosensor [72]. Furthermore, due to their high fluorescence intensity, photostability and long lifetimes than conventional organic fluorophores, QDs were also involved in the design of molecular beacons.

Bioanalytical systems using QDs as labels. QDs were first applied as bioanalytical labels in 1998, [73] an ultrasensitive bioanalytical system involving QDs was demonstrated for protein imaging. Since then, QDs have been widely applied in various bioanalytical methodologies, e.g., the enzyme-linked immunosorbent assay (ELISA), fluorescence resonance energy transfer (FRET) assay and cell tracking.

QDs can be simply used as fluorescent labels in immunosensors to quantify the biological targets through directly measuring the presence and/or the intensity of the fluorescence. CdTe/SiO2 core-shell structured nanoparticles were labeled with prostate-specific antigen (PSA) detection antibodies for PSA detection. And the fluorescent signals were measured after the specific recognition between PSA and the QD labeled antibodies followed by a magnetic separation to remove unbound QDs. This system represents a typical fluorescent biosensor detecting the presence of QD labels [74]. This strategy design shows great flexibility for employing a variety of fluorescence detectors, e.g., fluorescence spectrometry and handheld UV lamp tests. Furthermore, such design enables not only target quantification but also imaging multiple targets with different QD labels [75].

Imaging. QDs have also demonstrated their applicability in biomedical imaging, which is an important tool in diagnosis, visualization, treatment and prognosis of diseases [76]. QDs emit in the visible and near infrared ray wavelengths, with high brightness and excellent photostability, suitable for morphological studies. A representative examples of such applications encapsulated QDs in virus-lick particles as theranostic platforms to image viral behavior in living cells [77]. Furthermore, multifunctional SV40 virus-like particles was constructed encapsulating QDs bearing peptides recognizing early, developmental, and late stages of atherosclerosis, respectively, in live mice [78].

3.2.2 Labeling type

One of essential challenges to apply QDs in biosensors is immobilization of target recognition biomolecules onto the surface of QDs via stable bonding. In this part, methods for preparing QD-biomolecule conjugates will be presented and modified QDs applied as labels in biosensors are also summarized. Roughly, there are 4 strategies to prepare the QD-based bioconjugates:

1. Direct binding: proteins and nucleic acids can be immobilized on the QDs surface directly through interactions between the thiol-groups or imidazole-groups with the metal component of QDs, e.g., alkylthiol terminated DNAs are linked with QDs surface directly via dative thiol bond [79].

2. Conjugation via ligands: QDs can be functionalized firstly by ligands, such as carboxyl groups, hydroxyl groups and amino-groups, then covalently bond to biomolecules [80].

3. Conjugation via functional shell: QDs are capped with silane shell [81] or copolymers [82], then bond to biomolecules by the functional groups on the outer shell.

4. Conjugation via specific biological affinity: some types of biological affinity can be used to bond the QDs with biomolecules strongly and specifically, such as biotin-streptavidin interaction [83].

3.3 Upconversion fluorescent materials

Upconversion fluorescent materials are emerging fluorescent nanoparticles that can convert low frequent exciting light into high frequent emitting light by absorbing two-photons or multi-photons. The luminous mechanism of upconversion nanoparticles (UCNPs) is anti-stokes, which is opposite with the most fluorescent materials, including organic fluorescent dyes, quantum dots, fluorescent proteins, metal complexes, etc [84]. Because of the distinctive luminescence mechanism, UCNPs have some unique advantages, which make up the disadvantage of above other dyes. Firstly, UCNPs have improved biological tissue penetration. Secondly, UCNPs can reduce light damage on biological samples. Thirdly, UCNPs can effectively avoid the disturbance of autofluorescence from biological samples. Therefore, UCNPs have wide applications in biosensing and imaging [85]. The sensing principle of UCNPs-based probe is widely based on fluorescence resonance energy transfer (FRET) between UCNPs (donor) and other down-conversion fluorophores (acceptor). The reaction between UCNPs-based probe and the target biomolecules will trigger or hinder the FRET process, causing the quenching and enhancement of fluorescence for detection.

4. Dyes in electrochemical biosensors

Electrochemical biosensor is capable of providing specific quantitative or semi-quantitative analytical information using electrochemical transduction elements, e.g., charge-transfer complexes. Low-cost, energy efficient, portable, easy fabrication, and real-time sensing are major advantages of electrochemical biosensing platforms. Among the electrical signal molecules, there are several types of dyes with electrochemical activity, and they will be introduced in this part. In addition, the electrochemical signals generation mechanisms are explained and applications of these dyes as labels in the biosensors are also displayed.

4.1 Organic dye molecules

Methylene blue (MB) is a kind of derivative of phenothiazine and widely used a redox indicator and electron transfer medium in electrochemical analysis. For a typical DNA detection using MB labels, the distance between the MB and the electrode surface is adjusted by the change of conformation of the DNA probes labeled with MB, so that the peak current or the change in impedance can indicate the presence of the target DNA and quantitative concentration. This method is simple and versatile, however easy to be influenced by the solution environment. Despite MB, other organic dyes, such as gentian violet, ethyl green, Hoechst 33258 are also utilized in the electrochemical works. They are not as popular as MB, but show good performance in some biosensors.

4.2 Organometallic complexes

Organometallic complexes consist of centrally located metal atoms or ions and completely or partly coordinated organic ligands. The organometallic complexes with transition metals have the advantages of strong redox signal, good chemical stability, low toxicity, and high structural flexibility. They interact with biomolecules via the Intermolecular interaction force and electrostatic interaction.

Ferrocene (Fc), is a yellow organometallic complexes with transition metal (Fe) and aromatic ligands (cyclopentadiene rings). Because Fc has two freely rotating cyclopentadiene rings, it can be labeled with the biomolecules, such as DNA via hydrophobic interactions. As an electrical signal molecule, in the combination of bio-receptors and target molecules, Fc generates electrical signals mainly by adjusting the distance between the Fc and the electrode surface to realize the change of electrical signal and achieve the purpose of detection.

K3[Fe(CN)6]/K4[Fe(CN)6], is a pair of dyes with bright red and yellow color, respectively. Mainly, they are used as electron transfer agents in amperometric biosensors, to replac the natural electron transfer agents of the enzymes. In the commercial blood glucose meters, the glucose in the blood reacts with glucose oxidase and K3[Fe(CN)6] fixed on the surface of the test strip to produce gluconic acid and K4[Fe(CN)6]. Applying a constant working voltage to the test strip, K4[Fe(CN)6] is oxidized to K3[Fe(CN)6], generating an oxidation current that is proportional to the glucose concentration.

4.3 Nanomaterials

4.3.1 Quantum dots

One of the most commonly used electrochemical biosensor is cadmium selenide (CdSe) QDs, which employ as electrical signal molecules for the labeling of nucleic acid strands [86]. The Pb^{2+} cleavage ribozyme sequence was modified on the surface of the magnetic beads, and designed an electrochemical biosensor for detecting Pb2+ by using rolling circle amplification reaction and a signal probe labeled with CdS QDs [87]. Based on Ni^{2+} cleavage ribozyme and CdSe QDs, The Ni^{2+} was detected and the detection limit was 6.67 nmol/L [88]. As electrical signal molecules, QDs have versatility and low background signal, which has great application prospects.

4.3.2 Graphene quantum dots

Graphene quantum dots (GQDs) are actually sheets of graphene with dimensions less than 100 nm with sp^2 hybridized honeycomb structures, and their shapes are mostly circular and elliptical, but square and hexagonal QDs are also available. Basically, GQDs are characterized as graphene-like, consisting of C, O, and H as well as carbonyl, carboxyl, hydroxyl, and epoxy groups. GQDs can bind to ssDNA through π-π interactions, but it has no such effect on double-stranded DNA. Park et al. used GQDs as electrical signal substances to detect the Hg2+ concentration by measuring the current generated during the electrochemical reduction of GQDs [89].

4.3.3 Metal–organic frameworks

Metal–organic frameworks (MOFs) are crystalline materials with an infinitely regular and infinitely expanding periodic network structure formed by the self-assembly of metal ions and organic ligands through coordination bonds, covalent bonds, and weak intermolecular bonds (π-π stacking, van der Waals forces, hydrogen bonding, and other electrostatic interactions, etc.) [90]. MOFs are nanomaterials with good stability, large porosity, and specific surface area that are of great interest in gas storage, drug delivery, and sensors. Due to the intrinsic peroxidase catalytic activity, MOFs can also be used in electrochemical biosensors. Xu et al. constructed a Pb^{2+} electrochemical biosensor based on the MOFs prepared based on Fe [91], and AgPt nanoparticles are employed to increase its electrical conductivity and electrocatalytic activity, and the obtained sensitivity approaches 0.032 pmol/L. However, even though MOFs have enzymatic activity to improve sensitivity, their synthesis process is very complicated, and the characterization of the modification process is also very critical, so it is not suitable for routine use.

5. Conclusion

Investigation and evaluation of dyes play a vital role in the process of introduction novel labels and their corresponding sensing methods, which signify opportunities for the development of biosensors. This chapter highlights the utilization of dyes as biosensing labels and some most important sensing mechanisms for biological, biotechnological, and biomedical applications. These designs and applications have been much attracted for in vivo and in vitro analysis due to their high sensitivity and selectivity, fast response, biocompatibility, etc. Further developments in novel synthetic approaches of functional nanomaterials and sensing strategies will accelerate the discovery of unique properties of dyes, which will further improve their applications towards future biosensing platforms.

Acknowledgements

The authors are grateful to ÅForsk Foundation (grant number, 20-280), Formas (grant number, 2019-01583), STINT (grant number, IB2020-8594) and I Bergh scholarship. Qilu young scholar program of Shandong University (grant number, 11500082063141) is also acknowledged for the financial support.

Conflict of interest

The authors declare no conflict of interest.

Author details

Hu Li[1,2], Yuanyuan Han[3], Haiyan Zhao[4], Hassan Jafri[5] and Bo Tian[6*]

1 Shandong Technology Centre of Nanodevices and Integration, School of Microelectronics, Shandong University, Jinan, China

2 Department of Materials Science and Engineering, Uppsala University, Uppsala, Sweden

3 College of Biology, Hunan University, Changsha, China

4 College of Science, Hebei University of Science and Technology, Shijiazhuang, China

5 Faculty of Engineering and Technology, Mirpur University of Science and Technology, Kashmir, Pakistan

6 Department of Biomedical Engineering, Central South University, Changsha, China

*Address all correspondence to: tianbo@csu.edu.cn

References

[1] Clark Jr. LC, Lyons C. ELECTRODE SYSTEMS FOR CONTINUOUS MONITORING IN CARDIOVASCULAR SURGERY. Annals of the New York Academy of Sciences. 1962;102:29-45. DOI:https://doi.org/10.1111/j.1749-6632.1962.tb13623.x

[2] Mosbach K, Danielsson B. An enzyme thermistor. Biochimica et Biophysica Acta (BBA) - Enzymology. 1974;364:140-5. DOI:https://doi.org/10.1016/0005-2744(74)90141-7

[3] Völkl K-P, Opitz N, Lübbers DW. Continuous measurement of concentrations of alcohol using a fluorescence-photometric enzymatic method. Fresenius' Zeitschrift für analytische Chemie. 1980;301:162-3. DOI:10.1007/BF00467800

[4] Guilbault GG. Determination of formaldehyde with an enzyme-coated piezoelectric crystal detector. Analytical Chemistry. 1983;55:1682-4. DOI:10.1021/ac00261a010

[5] Liedberg B, Nylander C, Lunström I. Surface plasmon resonance for gas detection and biosensing. Sensors and Actuators. 1983;4:299-304. DOI:https://doi.org/10.1016/0250-6874(83)85036-7

[6] Thevenot DR, Toth K, Durst RA, Wilson GS. Electrochemical biosensors: recommended definitions and classification. Pure and applied chemistry. 1999;71:2333-48.

[7] Turner, A.P.F., Karube, I., and Wilson GS, editors. Biosensors: Fundamentals and Applications. Oxford: University Press; 1987. 770 p. 2011. 1132 p.

[8] Doering WE, Nie S. Spectroscopic Tags Using Dye-Embedded Nanoparticles and Surface-Enhanced Raman Scattering. Analytical Chemistry. 2003;75:6171-6. DOI:10.1021/ac034672u

[9] Wu D-Y, Li J-F, Ren B, Tian Z-Q. Electrochemical surface-enhanced Raman spectroscopy of nanostructures. Chemical Society Reviews. 2008;37:1025-41. DOI:10.1039/B707872M

[10] Qian X, Peng X-H, Ansari DO, Yin-Goen Q, Chen GZ, Shin DM, et al. In vivo tumor targeting and spectroscopic detection with surface-enhanced Raman nanoparticle tags. Nature Biotechnology. 2008;26:83-90. DOI:10.1038/nbt1377

[11] Fleischmann M, Hendra PJ, McQuillan AJ. Raman spectra of pyridine adsorbed at a silver electrode. Chemical Physics Letters. 1974;26:163-6. DOI:https://doi.org/10.1016/0009-2614(74)85388-1

[12] Le Ru EC, Etchegoin PG, Meyer M. Enhancement factor distribution around a single surface-enhanced Raman scattering hot spot and its relation to single molecule detection. The Journal of Chemical Physics. 2006;125:204701. DOI:10.1063/1.2390694

[13] Zhao Y, Liu X, Lei DY, Chai Y. Effects of surface roughness of Ag thin films on surface-enhanced Raman spectroscopy of graphene: spatial nonlocality and physisorption strain. Nanoscale. 2014;6:1311-7. DOI:10.1039/C3NR05303B

[14] Su J-P, Lee Y-T, Lu S-Y, Lin JS. Chemical mechanism of surface-enhanced raman scattering spectrum of pyridine adsorbed on Ag cluster: Ab initio molecular dynamics approach. Journal of Computational Chemistry. 2013;34:2806-15. DOI:https://doi.org/10.1002/jcc.23464

[15] Zhang L, Mazouzi Y, Salmain M, Liedberg B, Boujday S. Antibody-Gold Nanoparticle Bioconjugates for Biosensors: Synthesis, Characterization and Selected Applications. Biosensors

and Bioelectronics. 2020;165:112370. DOI:https://doi.org/10.1016/j.bios.2020.112370

[16] Duan N, Shen M, Qi S, Wang W, Wu S, Wang Z. A SERS aptasensor for simultaneous multiple pathogens detection using gold decorated PDMS substrate. Spectrochimica Acta Part A: Molecular and Biomolecular Spectroscopy. 2020;230:118103. DOI:https://doi.org/10.1016/j.saa.2020.118103

[17] Panikar SS, Banu N, Haramati J, Gutierrez-Silerio GY, Bastidas-Ramirez BE, Tellez-Bañuelos MC, et al. Anti-fouling SERS-based immunosensor for point-of-care detection of the B7–H6 tumor biomarker in cervical cancer patient serum. Analytica Chimica Acta. 2020;1138:110-22. DOI:https://doi.org/10.1016/j.aca.2020.09.019

[18] Zhu A, Ali S, Xu Y, Ouyang Q, Chen Q. A SERS aptasensor based on AuNPs functionalized PDMS film for selective and sensitive detection of *Staphylococcus aureus*. Biosensors and Bioelectronics. 2021;172:112806. DOI:https://doi.org/10.1016/j.bios.2020.112806

[19] Quesada-González D, Merkoçi A. Nanoparticle-based lateral flow biosensors. Biosensors and Bioelectronics. 2015;73:47-63. DOI:https://doi.org/10.1016/j.bios.2015.05.050

[20] Bhatt N, Huang P-JJ, Dave N, Liu J. Dissociation and Degradation of Thiol-Modified DNA on Gold Nanoparticles in Aqueous and Organic Solvents. Langmuir. 2011;27:6132-7. DOI:10.1021/la200241d

[21] Elghanian R, Storhoff JJ, Mucic RC, Letsinger RL, Mirkin CA. Selective Colorimetric Detection of Polynucleotides Based on the Distance-Dependent Optical Properties of Gold Nanoparticles. Science. 1997;277:1078 LP – 1081. DOI:10.1126/science.277.5329.1078

[22] Liu G, Lu M, Huang X, Li T, Xu D. Application of Gold-Nanoparticle Colorimetric Sensing to Rapid Food Safety Screening. Sensors (Basel, Switzerland). 2018;18:4166. DOI:10.3390/s18124166

[23] Huang J, Shangguan J, Guo Q, Ma W, Wang H, Jia R, et al. Colorimetric and fluorescent dual-mode detection of microRNA based on duplex-specific nuclease assisted gold nanoparticle amplification. Analyst. 2019;144:4917-24. DOI:10.1039/C9AN01013K

[24] Zagorovsky K, Chan WCW. A Plasmonic DNAzyme Strategy for Point-of-Care Genetic Detection of Infectious Pathogens. Angewandte Chemie International Edition. 2013;52:3168-71. DOI:https://doi.org/10.1002/anie.201208715

[25] Li J, Liu Q, Xi H, Wei X, Chen Z. Y-Shaped DNA Duplex Structure-Triggered Gold Nanoparticle Dimers for Ultrasensitive Colorimetric Detection of Nucleic Acid with the Dark-Field Microscope. Analytical Chemistry. 2017;89:12850-6. DOI:10.1021/acs.analchem.7b03391

[26] Posthuma-Trumpie GA, Wichers JH, Koets M, Berendsen LBJM, van Amerongen A. Amorphous carbon nanoparticles: a versatile label for rapid diagnostic (immuno)assays. Analytical and bioanalytical chemistry. 2012;402:593-600. DOI:10.1007/s00216-011-5340-5

[27] Blažková M, Rauch P, Fukal L. Strip-based immunoassay for rapid detection of thiabendazole. Biosensors and Bioelectronics. 2010;25:2122-8. DOI:https://doi.org/10.1016/j.bios.2010.02.011

[28] Yang Y, Li M, Tang A, Liu Y, Li Z, Fu S. Preparation of Covalent and Solvent-resistance Colored Latex Particles and Its Application on Cotton Fabric. Fibers and Polymers. 2020;21:1685-93. DOI:10.1007/s12221-020-9990-9

[29] Rembaum A, Dreyer WJ. Immunomicrospheres: reagents for cell labeling and separation. Science. 1980;208:364 LP – 368. DOI:10.1126/science.6768131

[30] Garcia VS, Guerrero SA, Gugliotta LM, Gonzalez VDG. A lateral flow immunoassay based on colored latex particles for detection of canine visceral leishmaniasis. Acta Tropica. 2020;212:105643. DOI:https://doi.org/10.1016/j.actatropica.2020.105643

[31] Matsumura Y, Enomoto Y, Takahashi M, Maenosono S. Metal (Au, Pt) Nanoparticle–Latex Nanocomposites as Probes for Immunochromatographic Test Strips with Enhanced Sensitivity. ACS Applied Materials & Interfaces. 2018;10:31977-87. DOI:10.1021/acsami.8b11745

[32] Borisov SM, Wolfbeis OS. Optical Biosensors. Chemical Reviews. 2008;108:423-61. DOI:10.1021/cr068105t

[33] Shen J, Li Y, Gu H, Xia F, Zuo X. Recent Development of Sandwich Assay Based on the Nanobiotechnologies for Proteins, Nucleic Acids, Small Molecules, and Ions. Chemical Reviews. 2014;114:7631-77. DOI:10.1021/cr300248x

[34] Hurley LH, Reynolds VL, Swenson DH, Petzold GL, Scahill TA. Reaction of the antitumor antibiotic CC-1065 with DNA: structure of a DNA adduct with DNA sequence specificity. Science. 1984;226:843 LP – 844. DOI:10.1126/science.6494915

[35] Wang J, Liu B. Highly sensitive and selective detection of Hg2+ in aqueous solution with mercury-specific DNA and Sybr Green I. Chemical Communications. 2008;4759-61. DOI:10.1039/B806885B

[36] Bhasikuttan AC, Mohanty J, Pal H. Interaction of Malachite Green with Guanine-Rich Single-Stranded DNA: Preferential Binding to a G-Quadruplex. Angewandte Chemie International Edition. 2007;46:9305-7. DOI:https://doi.org/10.1002/anie.200703251

[37] Maniatis T, Jeffrey A, Kleid DG. Nucleotide sequence of the rightward operator of phage lambda. Proceedings of the National Academy of Sciences. 1975;72:1184 LP – 1188. DOI:10.1073/pnas.72.3.1184

[38] Li H, Liu J, Fang Y, Qin Y, Xu S, Liu Y, et al. G-quadruplex-based ultrasensitive and selective detection of histidine and cysteine. Biosensors and Bioelectronics. 2013;41:563-8. DOI:https://doi.org/10.1016/j.bios.2012.09.024

[39] Day HA, Pavlou P, Waller ZAE. i-Motif DNA: Structure, stability and targeting with ligands. Bioorganic & Medicinal Chemistry. 2014;22:4407-18. DOI:https://doi.org/10.1016/j.bmc.2014.05.047

[40] Tyagi S, Kramer FR. Molecular Beacons: Probes that Fluoresce upon Hybridization. Nature Biotechnology. 1996;14:303-8. DOI:10.1038/nbt0396-303

[41] Zeng R, Luo Z, Su L, Zhang L, Tang D, Niessner R, et al. Palindromic Molecular Beacon Based Z-Scheme BiOCl-Au-CdS Photoelectrochemical Biodetection. Analytical Chemistry. 2019;91:2447-54. DOI:10.1021/acs.analchem.8b05265

[42] Masuko M, Ohtani H, Ebata K, Shimadzu A. Optimization of

excimer-forming two-probe nucleic acid hybridization method with pyrene as a fluorophore. Nucleic acids research. 1998;26:5409—5416. DOI:10.1093/nar/26.23.5409

[43] Kolpashchikov DM. Binary Probes for Nucleic Acid Analysis. Chemical Reviews. 2010;110:4709-23. DOI:10.1021/cr900323b

[44] Svanvik N, Westman G, Wang D, Kubista M. Light-Up Probes: Thiazole Orange-Conjugated Peptide Nucleic Acid for Detection of Target Nucleic Acid in Homogeneous Solution. Analytical Biochemistry. 2000;281:26-35. DOI:https://doi.org/10.1006/abio.2000.4534

[45] Fang X, Cao Z, Beck T, Tan W. Molecular Aptamer for Real-Time Oncoprotein Platelet-Derived Growth Factor Monitoring by Fluorescence Anisotropy. Analytical Chemistry. 2001;73:5752-7. DOI:10.1021/ac010703e

[46] Jhaveri SD, Kirby R, Conrad R, Maglott EJ, Bowser M, Kennedy RT, et al. Designed Signaling Aptamers that Transduce Molecular Recognition to Changes in Fluorescence Intensity. Journal of the American Chemical Society. 2000;122:2469-73. DOI:10.1021/ja992393b

[47] Wang H, Wang Y, Jin J, Yang R. Gold Nanoparticle-Based Colorimetric and “Turn-On” Fluorescent Probe for Mercury(II) Ions in Aqueous Solution. Analytical Chemistry. 2008;80:9021-8. DOI:10.1021/ac801382k

[48] Guilbault GG, editors. Practical Fluorescence, 2nd ed. Boca Raton: CRC Press; 2020. 826 p. DOI:10.1201/9781003066514

[49] Alvarez-Pez JM, Ballesteros L, Talavera E, Yguerabide J. Fluorescein Excited-State Proton Exchange Reactions: Nanosecond Emission Kinetics and Correlation with Steady-State Fluorescence Intensity. The Journal of Physical Chemistry A. 2001;105:6320-32. DOI:10.1021/jp010372+

[50] Haugland RP, editors. Handbook of fluorescent probes and research products [Internet]. 9th editon, EU version. Eugene (Or.) : Molecular probes; 2002. 966 p.

[51] Song L, Hennink EJ, Young IT, Tanke HJ. Photobleaching kinetics of fluorescein in quantitative fluorescence microscopy. Biophysical journal. 1995;68:2588-600. DOI:10.1016/S0006-3495(95)80442-X

[52] Jiao G-S, Han JW, Burgess K. Syntheses of Regioisomerically Pure 5- or 6-Halogenated Fluoresceins. The Journal of Organic Chemistry. 2003;68:8264-7. DOI:10.1021/jo034724f

[53] Banks PR, Paquette DM. Comparison of Three Common Amine Reactive Fluorescent Probes Used for Conjugation to Biomolecules by Capillary Zone Electrophoresis. Bioconjugate Chemistry. 1995;6:447-58. DOI:10.1021/bc00034a015

[54] Ceresole. M. Production of new red coloring matter. 1888;377360.

[55] Poronik YM, Vygranenko K V, Gryko D, Gryko DT. Rhodols – synthesis, photophysical properties and applications as fluorescent probes. Chemical Society Reviews. 2019;48:5242-65. DOI:10.1039/C9CS00166B

[56] Chen X, Pradhan T, Wang F, Kim JS, Yoon J. Fluorescent Chemosensors Based on Spiroring-Opening of Xanthenes and Related Derivatives. Chemical Reviews. 2012;112:1910-56. DOI:10.1021/cr200201z

[57] Mishra A, Behera RK, Behera PK, Mishra BK, Behera GB. Cyanines during the 1990s: A Review. Chemical Reviews.

2000;100:1973-2012. DOI:10.1021/cr990402t

[58] Fabian J, Nakazumi H, Matsuoka M. Near-infrared absorbing dyes. Chemical Reviews. 1992;92:1197-226. DOI:10.1021/cr00014a003

[59] Zhu Z, Chao J, Yu H, Waggoner AS. Directly labeled DNA probes using fluorescent nucleotides with different length linkers. Nucleic Acids Research. 1994;22:3418-22. DOI:10.1093/nar/22.16.3418

[60] Ekimov A ~I., Onushchenko A ~*A. quantum* size effect in three-dimensional microscopic semiconductor crystals. Soviet Journal of Experimental and Theoretical Physics Letters. 1981;34:345.

[61] Kalyanasundaram K, Borgarello E, Duonghong D, Grätzel M. Cleavage of Water by Visible-Light Irradiation of Colloidal CdS Solutions; Inhibition of Photocorrosion by RuO2. Angewandte Chemie International Edition in English. 1981;20:987-8. DOI:https://doi.org/10.1002/anie.198109871

[62] Rossetti R, Nakahara S, Brus LE. Quantum size effects in the redox potentials, resonance Raman spectra, and electronic spectra of CdS crystallites in aqueous solution. The Journal of Chemical Physics. 1983;79:1086-8. DOI:10.1063/1.445834

[63] Murray CB, Norris DJ, Bawendi MG. Synthesis and characterization of nearly monodisperse CdE (E = sulfur, selenium, tellurium) semiconductor nanocrystallites. Journal of the American Chemical Society. 1993;115:8706-15. DOI:10.1021/ja00072a025

[64] Mal J, Nancharaiah Y V, van Hullebusch ED, Lens PNL. Metal chalcogenide quantum dots: biotechnological synthesis and applications. RSC Advances. 2016;6:41477-95. DOI:10.1039/C6RA08447H

[65] Zhu H, Song N, Lian T. Controlling Charge Separation and Recombination Rates in CdSe/ZnS Type I Core–Shell Quantum Dots by Shell Thicknesses. Journal of the American Chemical Society. 2010;132:15038-45. DOI:10.1021/ja106710m

[66] Chung Y-C, Yang C-H, Zheng H-W, Tsai P-S, Wang T-L. Synthesis and characterization of CdSxSe1−x alloy quantum dots with composition-dependent band gaps and paramagnetic properties. RSC Advances. 2018;8:30002-11. DOI:10.1039/C8RA06007J

[67] Li Y-F, Feng J, Sun H-B. Perovskite quantum dots for light-emitting devices. Nanoscale. 2019;11:19119-39. DOI:10.1039/C9NR06191F

[68] Chung S, Revia RA, Zhang M. Graphene Quantum Dots and Their Applications in Bioimaging, Biosensing, and Therapy. Advanced Materials. 2019;n/a:1904362. DOI:https://doi.org/10.1002/adma.201904362

[69] Medintz IL, Uyeda HT, Goldman ER, Mattoussi H. Quantum dot bioconjugates for imaging, labelling and sensing. Nature Materials. 2005;4:435-46. DOI:10.1038/nmat1390

[70] Cotta M*a. quantum* Dots and Their Applications: What Lies Ahead? ACS Applied Nano Materials. 2020;3:4920-4. DOI:10.1021/acsanm.0c01386

[71] Ma F, Li C, Zhang C. Development of quantum dot-based biosensors: principles and applications. Journal of Materials Chemistry B. 2018;6:6173-90. DOI:10.1039/C8TB01869C

[72] Chern M, Toufanian R, Dennis AM. Quantum dot to quantum dot Förster resonance energy transfer: engineering materials for visual color change

sensing. Analyst. 2020;145:5754-67. DOI:10.1039/D0AN00746C

[73] Chan WCW, Nie S. Quantum Dot Bioconjugates for Ultrasensitive Nonisotopic Detection. Science. 1998;281:2016 LP – 2018. DOI:10.1126/science.281.5385.2016

[74] Zhao Y, Gao W, Ge X, Li S, Du D, Yang H. CdTe@SiO2 signal reporters-based fluorescent immunosensor for quantitative detection of prostate specific antigen. Analytica Chimica Acta. 2019;1057:44-50. DOI:https://doi.org/10.1016/j.aca.2019.01.019

[75] Wu M, Zhang Z-L, Chen G, Wen C-Y, Wu L-L, Hu J, et al. Rapid and Quantitative Detection of Avian Influenza A(H7N9) Virions in Complex Matrices Based on Combined Magnetic Capture and Quantum Dot Labeling. Small. 2015;11:5280-8. DOI:https://doi.org/10.1002/smll.201403746

[76] Mateu MG. Assembly, Engineering and Applications of Virus-Based Protein Nanoparticles. Advances in experimental medicine and biology. 2016;940:83-120. DOI:10.1007/978-3-319-39196-0_5

[77] Li F, Zhang Z-P, Peng J, Cui Z-Q, Pang D-W, Li K, et al. Imaging viral behavior in Mammalian cells with self-assembled capsid-quantum-dot hybrid particles. Small (Weinheim an der Bergstrasse, Germany). 2009;5:718-26. DOI:10.1002/smll.200801303

[78] Sun X, Li W, Zhang X, Qi M, Zhang Z, Zhang X-E, et al. In Vivo Targeting and Imaging of Atherosclerosis Using Multifunctional Virus-Like Particles of Simian Virus 40. Nano letters. 2016;16:6164-71. DOI:10.1021/acs.nanolett.6b02386

[79] Banerjee A, Pons T, Lequeux N, Dubertret B. Quantum dots-DNA bioconjugates: synthesis to applications. Interface focus. 2016;6:20160064. DOI:10.1098/rsfs.2016.0064

[80] Medintz IL, Konnert JH, Clapp AR, Stanish I, Twigg ME, Mattoussi H, et al. A fluorescence resonance energy transfer-derived structure of a quantum dot-protein bioconjugate nanoassembly. Proceedings of the National Academy of Sciences of the United States of America. 2004;101:9612 LP – 9617. DOI:10.1073/pnas.0403343101

[81] Feng H, ten Hove JB, Zheng T, Velders AH, Sprakel J. All-Aqueous Synthesis of Silica-Encapsulated Quantum Dots with Functional Shells. European Journal of Inorganic Chemistry. 2017;2017:5152-7. DOI:https://doi.org/10.1002/ejic.201700886

[82] Palui G, Aldeek F, Wang W, Mattoussi H. Strategies for interfacing inorganic nanocrystals with biological systems based on polymer-coating. Chemical Society Reviews. 2015;44:193-227. DOI:10.1039/C4CS00124A

[83] Díaz-González M, de la Escosura-Muñiz A, Fernandez-Argüelles MT, García Alonso FJ, Costa-Fernandez JM. Quantum Dot Bioconjugates for Diagnostic Applications. Topics in Current Chemistry. 2020;378:35. DOI:10.1007/s41061-020-0296-6

[84] Zhou J, Liu Q, Feng W, Sun Y, Li F. Upconversion Luminescent Materials: Advances and Applications. Chemical Reviews. 2015;115:395-465. DOI:10.1021/cr400478f

[85] Wang F, Banerjee D, Liu Y, Chen X, Liu X. Upconversion nanoparticles in biological labeling, imaging, and therapy. Analyst. 2010;135:1839-54. DOI:10.1039/C0AN00144A

[86] Fan H, Chang Z, Xing R, Chen M, Wang Q, He P, et al. An Electrochemical Aptasensor for Detection of Thrombin

based on Target Protein-induced Strand Displacement. Electroanalysis. 2008;20:2113-7. DOI:https://doi.org/10.1002/elan.200804281

[87] Tang S, Lu W, Gu F, Tong P, Yan Z, Zhang L. A novel electrochemical sensor for lead ion based on cascade DNA and quantum dots amplification. Electrochimica Acta. 2014;134:1-7. DOI:https://doi.org/10.1016/j.electacta.2014.04.021

[88] Yang Y, Yuan Z, Liu X-P, Liu Q, Mao C-J, Niu H-L, et al. Electrochemical biosensor for Ni2+ detection based on a DNAzyme-CdSe nanocomposite. Biosensors and Bioelectronics. 2016;77:13-8. DOI:https://doi.org/10.1016/j.bios.2015.09.014

[89] Park H, Hwang S-J, Kim K. An electrochemical detection of Hg2+ ion using graphene oxide as an electrochemically active indicator. Electrochemistry Communications. 2012;24:100-3. DOI:10.1016/j.elecom.2012.08.027

[90] Xu W, Zhou X, Gao J, Xue S, Zhao J. Label-free and enzyme-free strategy for sensitive electrochemical lead aptasensor by using metal-organic frameworks loaded with AgPt nanoparticles as signal probes and electrocatalytic enhancers. Electrochimica Acta. 2017;251:25-31. DOI:https://doi.org/10.1016/j.electacta.2017.08.046

[91] Rogez G, Massobrio C, Rabu P, Drillon M. Layered hydroxide hybrid nanostructures: a route to multifunctionality. Chemical Society Reviews. 2011;40:1031-58. DOI:10.1039/C0CS00159G

Chapter 4

Natural Dyes: From Cotton Fabrics to Solar Cells

Indriana Kartini and Adhi Dwi Hatmanto

Abstract

This article will discuss natural dyes' role, from colouring the cotton fabrics with some functionality to harvesting sunlight in the dye-sensitized solar cells. Natural dye colourants are identical to the low light- and wash-fastness. Therefore, an approach to improving the colourant's physical properties is necessary. Colouring steps employing silica nanosol and chitosan will be presented. The first part will be these multifunctional natural dye coatings on cotton fabrics. Then, functionality such as hydrophobic surfaces natural dyed cotton fabrics will be discussed. Natural dyes are also potential for electronic application, such as solar cells. So, the second part will present natural dyes as the photosensitizers for solar cells. The dyes are adsorbed on a semiconductor oxide surface, such as TiO_2 as the photoanode. Electrochemical study to explore natural dyes' potential as sensitizer will be discussed, for example, natural dyes for *Batik*. Ideas in improving solar cell efficiency will be discussed by altering the photoanode's morphology. The ideas to couple the natural dyes with an organic–inorganic hybrid of perovskite and carbon dots are then envisaged.

Keywords: natural dyes, cotton fabrics, hydrophobic, multifunctional textiles, dye-sensitized solar cells

1. Introduction

Technology is a means to achieve enhanced goals towards advancing human civilization, as is textile dyeing technology. Dyeing is an integral part of the wet textile processing process, which involves massive amounts of chemicals, both in type and quantity. Recently, the development of the concept of eco-fashion or sustainable textiles has led to the development of dyeing technology using natural dyes that care about aspects of water pollution, the sustainability of raw materials and processed products, biodegradability and other environmentally friendly attributes [1]. Human awareness of a healthy environment has revived interest in products that use natural dyes.

Eco-fashion or Sustainable Fashion as a trend against fast fashion is part of a developing design philosophy to create a system that can support and counteract the impact of human activities on the environment. The focus of eco-fashion is not only on the aspects of the materials used and the environment affected by it but also on the wearer's health and the durability of the clothes. An example is the use of natural pesticide-free materials, the use of materials that can be recycled, clothes that are made to last longer and are not easily damaged, to cover the welfare guarantee for fashion workers. Conventional clothing production is known to involve many resources and produce hazardous waste for the environment. Three criteria attached to environmentally friendly textile creation products include less toxic chemicals, less land or water, and reduction of greenhouse gases. The advantages of nano-sized materials promise exploration opportunities for new technologies with achievements beyond those achieved in computers and biotechnology in recent decades.

The advantage of using natural dyes lies in the smoothness and softness of the colour. This product is highly valued and maintained because it reflects the beauty, prestige, and cultural structures whose existence cannot be replaced by synthetic dyes. However, despite the advantages of natural dyes, several shortcomings of natural dyes have made *Batik* (tradisitional Indonesia fabrics) craftsmen still reluctant to change their *Batik* dyes from synthetic dyes to natural dyes. Among them are the high price, limited availability, long manufacturing process, and low colour resistant to light or washing.

Natural dyes that are currently often used for *Batik* production besides indigo blue are *Tingi* (*Ceriops tagal*) natural dyes. This dye is obtained from the extraction of the bark of the *Tingi* tree, a type of mangrove plant, which has a high tannin content and is used as a dye for *Batik* and tanners. This dye gives the distinctive brown colour of *Batik*. Like most other natural dyes, *Tingi* natural dyes also have a low degree of fastness to washing. So it requires treatment to increase its fastness to washing. Efforts to increase the colour resistance to washing of cotton fabrics coloured by *Tingi* natural dyes will be discussed in the next section. Afterward, works to attach hydrophobic functionality to result in a multifunctional textiles will be described.

Nanotechnology is a technology related to materials or systems at the nanometer scale (1 nm = 10^{-9} m). Unusual changes, which cannot be predicted using classical mechanical models, will be obtained at the nanometer scale, such as changes to electronic properties, mechanical properties, magnetic properties, optical properties and chemical reactivity. The potential for a revival of natural dyes can occur through treaties with nanotechnology.

One of the breakthroughs in photovoltaic technology was the photovoltaic cell's invention based on the photoelectrochemical concept employing nanomaterial by a group of Swiss researchers [2], which became popular as dye-sensitized solar cells (DSSC). The solar cell is composed of a thin layer of semiconductor material, such as titanium dioxide or titania (TiO_2), which has a porous structure, a complex ruthenium (Ru) compound as a sensitizer, and an electrolyte system for the redox pair of iodine compounds. The ruthenium dye complex has a role in absorbing solar radiation, which will generate the dye's electron system so that it flows into the semiconductor material and is connected to a circuit to generate an electric current. The excited electrons from the dye are immediately replaced by the electrons produced from the electrolyte redox pair system, I^-/I_3^-. The natural mechanisms of photosynthesis inspire technology to harvest and use continuous sunlight as a source of energy for all life on earth. Harvesting sunlight ultimately requires the sensitizer to have an absorption character like a black body. So far, ruthenium complex dye as a DSSC photosensitizer has produced a conversion efficiency of ~10% [2]. However, ruthenium is not environmentally friendly. Therefore environmentally friendly sensitizers need to be sought. This environmental demand raises the potential of natural dyes as solar cell sensitizers.

Natural sources of natural dyes for sensitizers are directed to plants that have no potential as a food source and have a large percentage of active colouring agents. Several natural dyes that can be used as solar cell sensitizers have been identified to contain tannins, anthocyanins, betalains, flavonoids, and carotenoids [3]. *Batik*'s natural dyes used for production, mostly are rich of tannins. Potential of this natural dyes will be explored in the fourth section. Finally, some ideas to improve the performance of the natural dyes solar cell will be envisaged in the concluding remarks.

2. Improving the wash-fastness of the natural dyed cotton fabrics

The wood of the *tingi* tree (**Figure 1**) is usually used as firewood. The bark is used as a dye for *Batik* and tanners because of its high tannin content. According to Kasmudjiastuti [4] the tannin content in the bark reaches as high as 26%. The *tingi* bark gives a reddish brown colour with a large enough tannin content. The availability of *tingi* bark as a raw material is very abundant in Indonesia. According to Nazir [5] tannins from *Tingi* dyes fall into the category of condensation tannins, with 26% more tannins than other woody plants such as Avaram, Hemlock, Oak, and Chestnut. Kasmudjiastuti [4] characterised the extract of *tingi* tree wood, resulting in that *tingi* wood contains 70.91% of tannins which are included in procyadinin condensation tannins. However, natural dyes derived from plant extraction have a weakness in their fastness resistance to washing processes and exposure to light. Modification of the dye composition can increase the dye fastness [6].

Dipping the dyed cotton into silica nanosols using the sol–gel method can improve the fastness resistance of a synthetic dye of malachite green b (MG) on cotton fabrics [7, 8]. The hydrogen interaction that occurs between the hydroxyl groups on the cellulose fibers and the hydroxyl groups from the silica sol probably made the silica-MG nanosols to be firmly coated on cotton fabrics. The thin silicon dioxide layer forms a layer that is resistant to heat, light, chemical processes and microbial attack. The oxide thin layer can improve the properties of mechanical strength and resistance to abrasion [7].

The silica nanosol was prepared using the sol gel method with tetraethylortosilicate (TEOS) as a precursor for Si. This process was carried out in an acidic solution of pH 3-4 using HCl as the catalyst and pH regulator [8]. **Figure 2** showed UV–Vis spectra of the *Tingi* extract in water and the mixture of silica nanosol and *Tingi* extract in volume ratio of 1:4 and 1:40. The maximum absorbance of the natural dye is at 473 nm and did not show any shifting after mixing with nanosol silica indicating no structure changes in the dye and the sols. The infrared spectra of the corresponding dried-powder of the mixture dye sols confirmed this, as implied in **Figure 3**. The more the dyes in the mixture sols, the weaker the peaks for Si-O-Si, at around 1080 cm^{-1}.

The dyeing process on the fabric was done by using the dip coating method, which is the direct immersion of the cloth in a solution mixture of silica sol and the

Figure 1.
The Tingi *tree (left) and its corresponding bark for the natural dye's resource (right).*

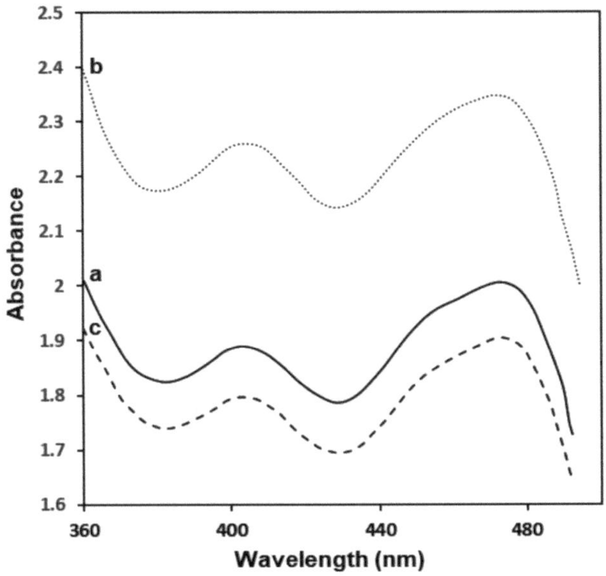

Figure 2.
Electronic spectra of: a. Tingi *extract, and silica sol-*Tingi *extract of: b. 1:4, c. 1:40 by volume.*

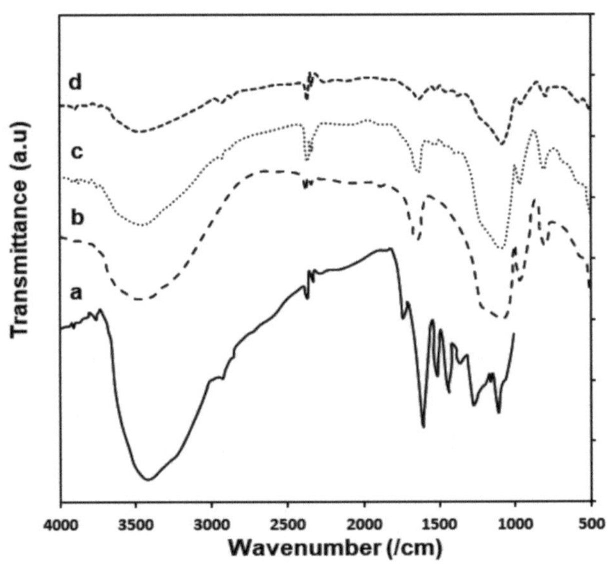

Figure 3.
Infrared spectra of: a. Tingi *extract powder, dried-powder of: b. silica sol, and silica sol-extract* Tingi *of: c. 1:4, d. 1:40.*

dye extract. The variation of the volume ratio of the silica-dye sol was 1:40; 1:8; 1:5; and 1:4 with a total volume of 50 mL. The photos of the dyeing products are displayed in **Figure 4**. The strong dark brown colours are the dominant colour. The colour strength changed as the sols composition changed, with the strongest observed for fabric coloured by 1:4 mixture sols of Si-*Tingi* extract. At other compositions, the colour are almost the same.

The wash-fastness of the dyed fabrics were tested by immersing the testing samples in 1% SDS (sodium dodecyl sulphate) solution (in water) at room

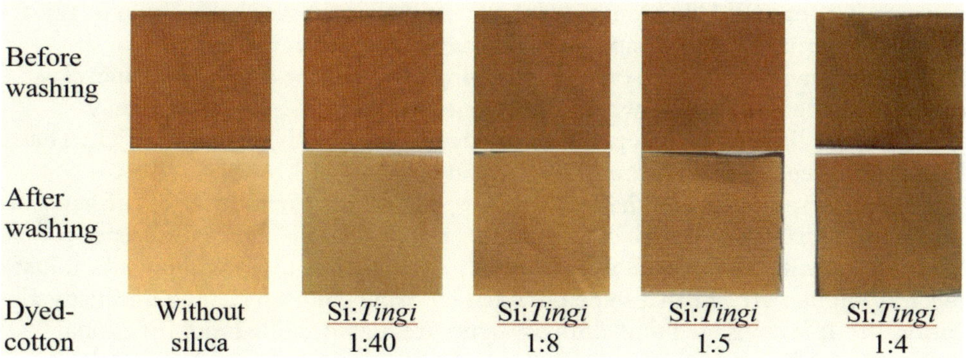

Figure 4.
Dyed cotton fabrics before and after washing under indoor illumination.

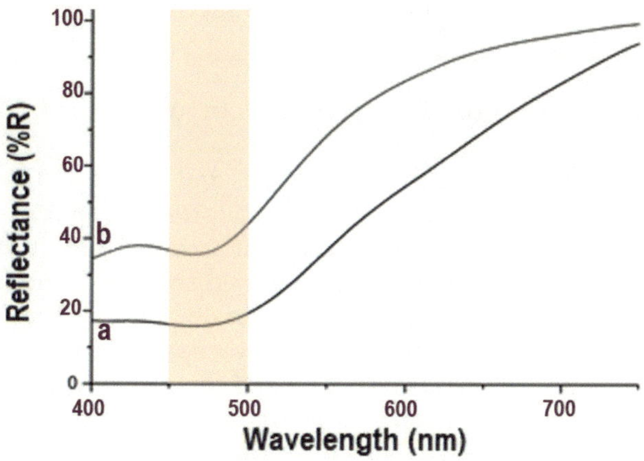

Figure 5.
Reflectance spectra of dyed-cotton without silica coating: a. before washing, b. after washing.

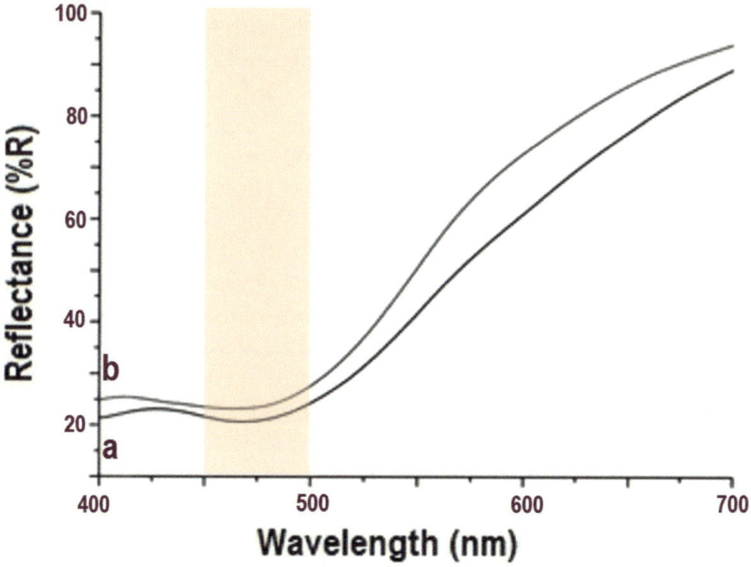

Figure 6.
Reflectance spectra of dyed-cotton with 1:4 silica coating: a. before washing, b. after washing.

temperature for 1 h [7, 8]. Compared to cotton cloth without the addition of silica nanosol, the mixture composition of silica nanosol-dye can increase the wash-fastness resistance of the dye over the washing process. The SiO_2-*Tingi* nanosol ratio of 1: 4 gave the best results, where the colour after the washing process only changed very little when compared to the dyed cotton without nanosol SiO_2. The leaching degree calculated from the reflectance data was 3.18%.

Figures 5 and **6** showed the reflectance spectra to confirm the effect of nanosol silica in the mixture of dye sols. It can be seen that the reflectance difference for fabrics dyed with silica nanosols is relatively smaller than those without silica. Just recently, similar effect can also be obtained by using chitosan coating on the dyed-cotton [9]. It is envisaged that chitosan structure may provide more functional groups for hydrogen bonding with either cellulose of the cotton fabrics or the dye (represented by procyanidin as the active dye for the *Tingi* extract). Therefore, the mixture of chitosan and dye solutions resulted in lower leaching degree to SDS than that of the dye itself. Leaching degree as low as 6.24% has been achieved for dyeing process using a mixture of chitosan and *Tingi* extract [9].

3. Hydrophobic surfaces on natural dyed cotton fabrics

Batik is a work of art with distinctive patterns and motifs on the fabric. The *Batik* cloth used is a cloth that has gone through a pre-treatment preparation process in the textile industry. The pre-treatment process gives a different character to the *Batik* cloth, the *Batik* fabrics commonly used are calico, cotton, and mori. *Batik* fabrics, which are natural textiles, are generally made of cellulose (cotton) and protein (silk) so they are considered more susceptible to microbial attack than synthetic fibers because the porous structure and the constituent polymers are hydrophilic so they are easy to absorb moisture [10]. Fabric surface engineering of *Batik* material needs to be done so that the fabric surface becomes hydrophobic and indirectly provides antimicrobial properties. Topographical engineering of micro-structure and chemical properties on the surface of the fabric was carried out using the sol–gel method.

So far, the surface preparation of hydrophobic fabrics has been done using fluorocarbons which are known to be compounds with low surface energy. Hayn et al. [11] conducted coating of fluorosilane compound (FS) on a nylon-cotton blend fabric resulting in a water contact angle of 148°. However, the use of fluorinated compounds which are commonly used as hydrophobic agents is now starting to be abandoned due to adverse effects such as pollution caused by high toxicity, bioaccumulation in living things and the costs used are also relatively expensive [12]. This has led to research using non-fluorine compounds which are more environmentally friendly. One of them is the compounds of the alkylsilane group which are known to have low surface energy, for example trimethylchlorosilanes (TMCS), octadecyltriclorosilanes (ODTCS), cetyltrimethoxysilanes (CTMS), and hexadecyltrimethoxysilanes (HDTMS) [13]. Here, we used HDTMS as the hydrophobic agent.

Three types of fabrics commonly used for *Batik* are cotton, *mori* and calico. The three types of clothes are batik fabrics which are differentiated based on the fabrication process. Calico cloth is a cellulose-based cloth that does not go through a pre-treatment process, while cotton and *mori* fabrics go through a pre-treatment process. Therefore, there are differences in fabric properties that will affect the interaction with silica nanosols and HDTMS. **Figure** 7 shows the water contact angle obtained from the surface of the three types of *Batik* common fabrics. Cotton and calico clothes resulted in similar basic water contact angle, so similar hydrophobicity. Therefore, for further testing using *Tingi* dyed fabrics, we used cotton.

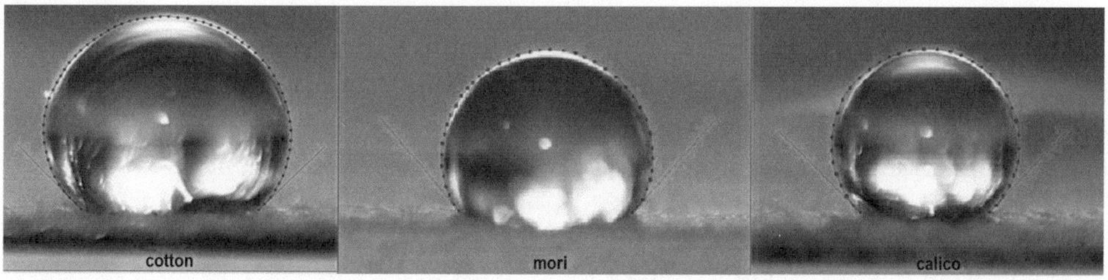

Figure 7.
The water contact angle on different types of Batik's *fabrics: cotton 135.8°, mori 133.9°, and calico 136.2°.*

Figure 8.
The water contact angle on cotton fabrics dyed by: a. Tingi *extract (120.1°), b.* Tingi*-silica nanosol mixture (134.7°),* C. tingi *and silica nanosol layer by layer (114.7°).*

Figure 8 displays the water contact angle of *Tingi*-dyed fabrics with and without silica nanosols coated by HDTMS. The mixture nanosol coated cloth showed the best hydrophobicity properties with the greatest water contact angle value of 134.7°, while the fabric coated layer by layer gave the lowest hydrophobicity. This could be due to the weak interaction between HDTMS and the dye molecules. The layer by layer coatings on cotton fabrics were performed in the sequence of silica nanosol, the dye, and the HDTMS.

Our recent results for chitosan coating mixture have shown improved water contact angle after leaching test using natural detergent (*Sapindus rarak*). Saponin in the *Sapindus rarak* which also classified as the low surface energy compound is presumably responsible for this enhanced hydrophobicity. A ten percent improvement was achieved for the fabrics dyed by a mixture of chitosan-*Tingi* extract dye, resulted in water contact angle of 107.83° [9]. Further studies are still required to explore the potential of *Sapindus rarak* as the co-hydrophobic agent to obtain a hydrophobic *Batik* fabrics.

4. Natural dyes for dye-sensitized solar cells: *Batik* and Algae's extract

A dye-sensitized solar cell (DSSC) is one promising alternative to conventional semiconductor silicon-based solar cells due to its low-cost and moderate efficiency. DSSC is typically constructed of TiO_2 (titania) nanoparticles film sensitized with a monolayer of dye molecules as the photoanode. Upon light illumination, the photo-excited dye molecules inject the electrons. Then, the electrons transport through the photoanode to the counter electrode (e.g., fluorine-doped tin oxide (FTO)). These electrons are collected at the counter electrode through an external load and further shuttled back to the oxidized dye molecules via redox reactions of I^-/I_3^- redox couple in the electrolyte. The dye molecules are critical to the overall device performance since they determine the amount of solar energy absorbed by the device. The efficiencies of the sensitizers are related to some essential criteria. The HOMO potential of the dye should be sufficiently positive compared to the electrolyte redox potential for efficient dye regeneration. The dye's LUMO potential should be negative enough to match the potential of the conduction band edge of the TiO_2. Its orbitals should be located at the acceptor part of the dye to provide efficient electron injection. The common dyes in DSSCs are based on ruthenium metal–ligand complexes (e.g. N3 and N719 dyes). However, the limited availability of ruthenium and the low stability of ruthenium-based dyes could hinder the commercialization of DSSCs. On the other hand, natural dyes are promising sensitizers for DSSC application because of their high extinction coefficient and variable chemical structures for strong and broad absorption of solar energy. In addition to the consideration of environmental aspects, natural dyes can also be extracted easily through water, methanol, or ethanol extraction process directly from the bark, roots, flowers, or leaves, so that they are cost-effective in comparison to the manufactured Ru dyes [14, 15].

Some natural dyes, including dyes extracted from the bark of *Tingi* (*Ceriops tagal, CT*) and *Tegeran* (*Maclura cochinensis, MC*), the dried fruit of *Jalawe* (*Terminalia bellirica(gaertn)roxb, TB*), as well as the leaves of Indigo (*Indigofera tinctoria, IT*), are commonly used in the production of *Batik* (**Figure 9**), a technique of wax-resist dyeing applied to whole cloth originated from Java Island in Indonesia. The bark of *CT* is silvery-grey to orangeish-brown, smooth with occasional pustular lenticels, containing 23-40% tannin. Like *CT*, the smooth, lenticellate, and yellowish-brown bark of *MC* contains a high amount of tannin. The dried fruit of *TB* is yellowish-brown with flavonoids, sterols, and tannins content [16], while the *IT* dye contains 2,2'-Bis(2,3-dihydro-3-oxoindolyliden), known as Indigotin, with a dark blue colour. Since all of those "Batik" natural dyes are able to absorb light, the possibility of using them as photosensitizers for DSSC will then become interesting and important to be further investigated. Considering that the energy level of the photosensitizers will strongly affect the electron transport in DSSC, in this study, the absorption spectra and electrochemical properties of the *Batik* natural dyes were

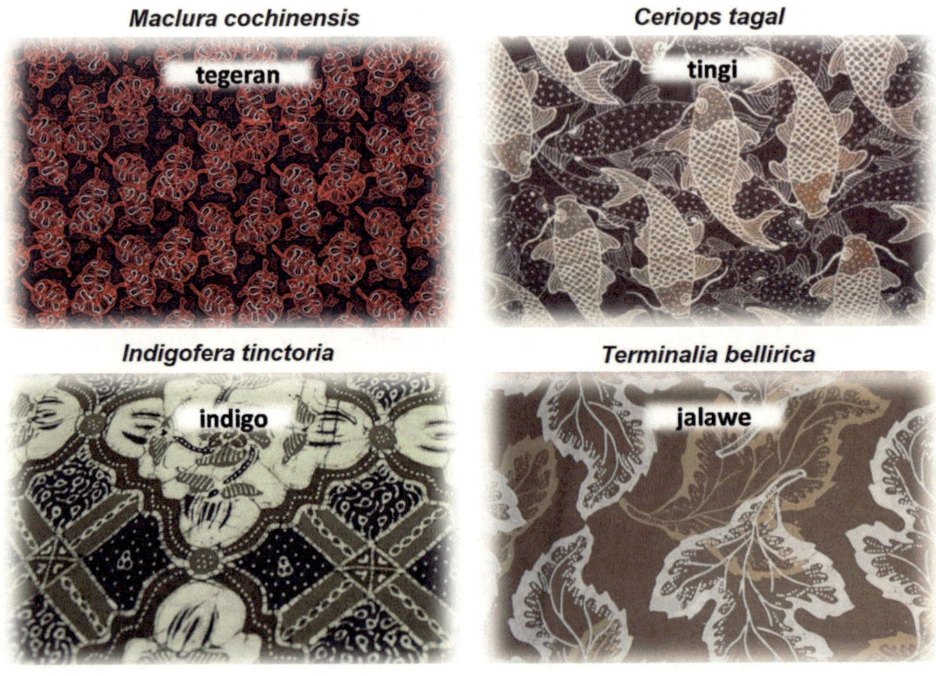

Figure 9.
Batik *with some Indonesian natural dyes.*

presented and discussed. Both data were used to construct the energy of the highest occupied molecular orbitals (HOMO) and lowest unoccupied molecular orbitals (LUMO) of the corresponding natural dyes to reveal their potential as a light harvester for DSSC.

The construction of schematic energy diagrams in DSSC requires some information regarding the HOMO and LUMO energy levels of the photosensitizer that are determined from its absorption spectra and electrochemical properties. The electronic spectra of the *Batik* natural dye extracts were determined using the UV–Vis spectrophotometry method in the range of 300 to 800 nm, as shown in **Figure 10**. The bark of *MC* and *CT*, as well as the dried fruit of *TB*, were extracted by heating in distilled water, while the *IT* dye was prepared by dissolving a commercial Indigo paste directly in ethanol. The dye extracted from the bark of *MC* shows a single absorption at 490 nm, while several absorptions in the range of 450-500 nm (with the highest peak at 482 nm) were observed from the dye extracted from the bark of *CT*. Both dyes extracted from *IT* and *TB* show a single absorption peak respectively at 665 and 370 nm. The energy band gap of materials was then determined by using the absorption edge of the spectrum. The absorption edge of *MC*, *CT*, *IT*, and *TB* were obtained at observed at 538, 540, 718, and 403 nm, respectively, which attributed to the bandgap energy (E_{gap}) of 2.305, 2.297, 1.729, and 3.078 eV. These E_{gap} values, together with the E_{HOMO} (determined from cyclic voltammetry analysis), were then used to calculate the LUMO energy level.

The electrochemical properties of all *Batik* natural dyes were studied by cyclic voltammetry method using Pt as the working electrode, Pt-wire as the auxiliary electrode, and Ag/AgCI as the reference electrode, with the addition of I^-/I_3^- redox couple as supporting electrolyte. The cyclic voltammograms of all four dyes are shown in **Figure 11**. All cyclic voltammograms results show combined peaks characteristic to

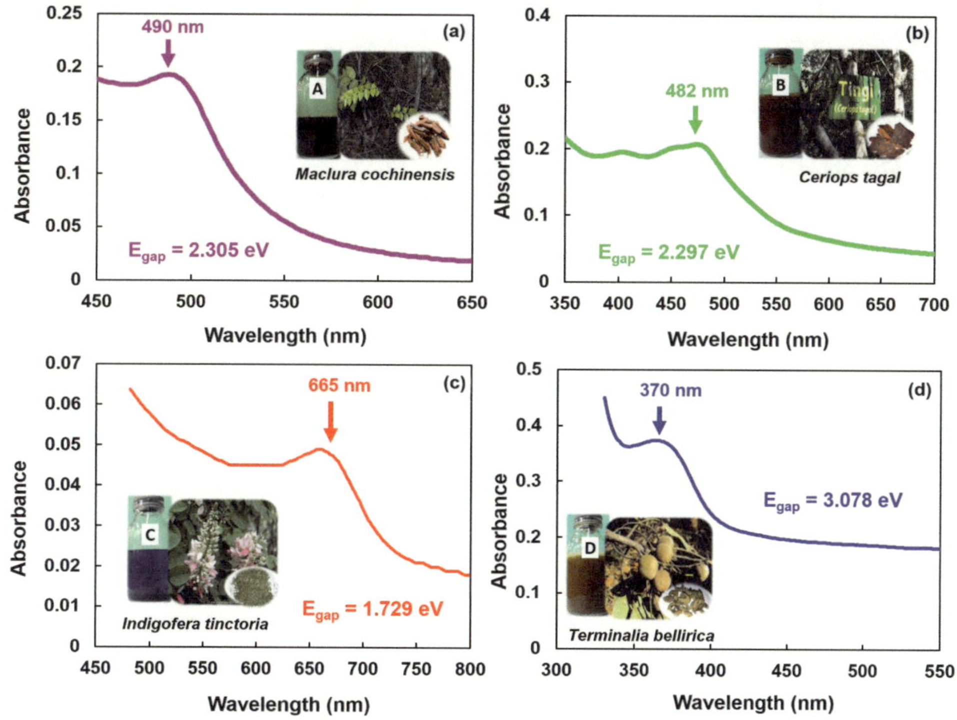

Figure 10.
UV–Vis absorption spectra of four Batik natural dyes: (a) Maclura cochinensis *(MC), (b)* Ceriops tagal *(CT), (c)* Indigofera tinctoria *(IT), (d)* Terminalia bellirica *(TB).*

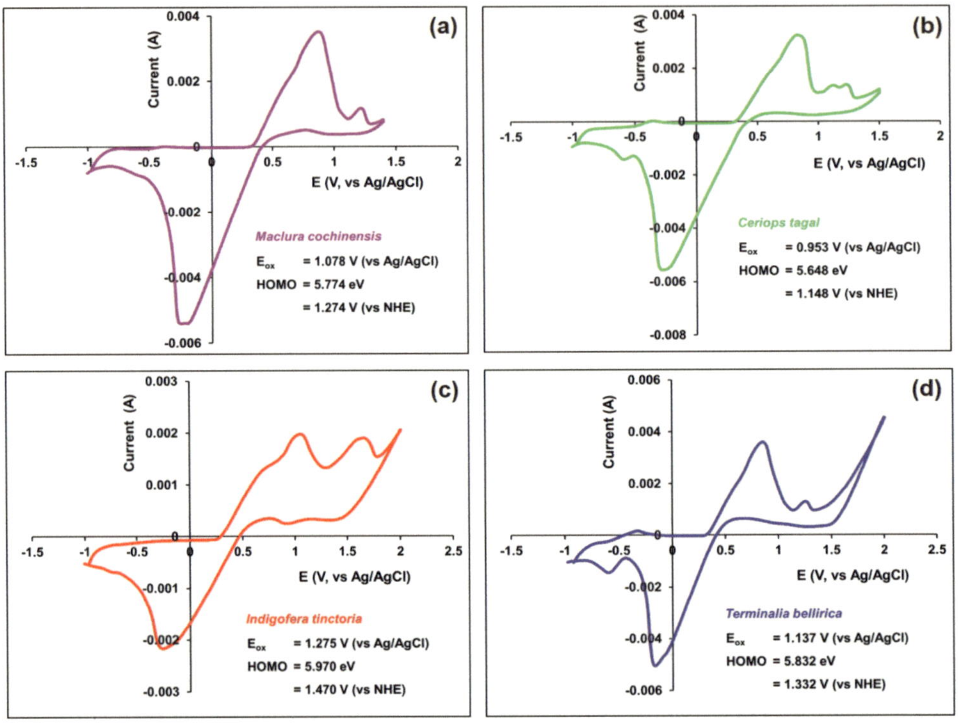

Figure 11.
Cyclic voltammograms of four Batik natural dyes: (a) Maclura cochinensis *(MC), (b)* Ceriops tagal *(CT), (c)* Indigofera tinctoria *(IT), (d)* Terminalia bellirica *(TB).*

oxidation and reduction potential of the reference electrolyte and the natural dyes. The HOMO energy level of the dyes was then calculated from the onset anodic potential of the cyclic voltammograms. The onset anodic potential (E_{ox}) is a cross-section of the baseline and the oxidation peak of the dye [17]. $Fe(CN)_6^{4-}/Fe(CN)_6^{3-}$ redox couple was used as an external standard to calculate the E_{HOMO} of the natural dyes. The onset anodic potential of *MC*, *CT*, *IT*, and *TB* were observed respectively at 1.078, 0.953, 1.275, and 1.137 V, which are attributable to the E_{HOMO} of 1.274, 1.148, 1.470, and 1.332 V (vs NHE), respectively. The E_{LUMO} was then calculated based on the bandgap energy and the E_{HOMO} of the dyes. They are −1.031, −1.149, −0.259, and − 1.746 V (vs NHE), respectively for *MC*, *CT*, *IT*, and *TB*. The half-wave redox potential ($E_{p/2}$) of I^-/I_3^- redox couple that was used as supporting electrolyte was observed at around 0.478 V vs. Ag/AgCl or 0.701 V vs. NHE. The values of E_{gap}, E_{HOMO}, and E_{LUMO} of all *Batik* natural dyes were summarized in **Table 1**.

Figure 12 shows a schematic energy level diagram of DSSC using *Batik* natural dyes as photosensitizer and I^-/I_3^- a couple as redox electrolyte. All the HOMO levels of the dyes are sufficiently more positive than the half-wave redox potential

Dyes	Absorption Edge (nm)	E_{gap} (V)	HOMO (V vs. NHE)	LUMO (V vs. NHE)
MC	538	2.305	1.274	−1.031
CT	540	2.297	1.148	−1.149
IT	718	1.729	1.470	−0.259
TB	403	3.078	1.332	−1.746

Table 1.
The values of E_{gap}, E_{HOMO}, and E_{LUMO} of the four Batik *natural dyes.*

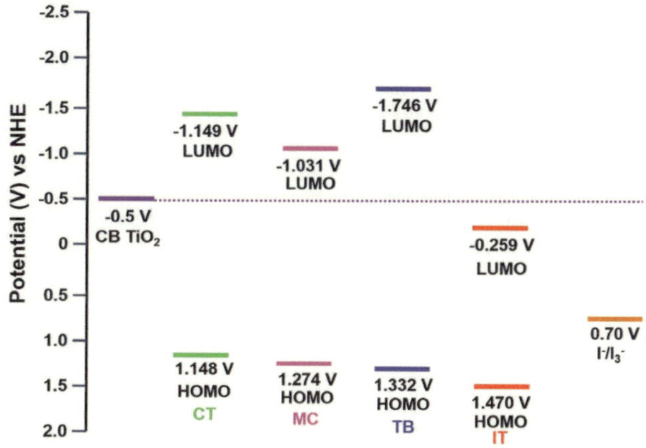

Figure 12.
Schematic energy level diagram of DSSC using Batik *natural dyes as photosensitizer and* I^-/I_3^- *couple as redox electrolyte.*

of I^-/I_3^- couple, suggesting an efficient regeneration of the oxidized dye by the I^-/I_3^- redox couple as the hole transport material. Meanwhile, the LUMO level of the dyes is sufficiently more negative than the conduction band edge of the TiO_2 (E_{CB}), except for IT, which ensure the necessary driving force for electron injection from the excited state of the dye into the conduction band of TiO_2 semiconductor [18, 19]. Therefore, in this DSSC system, we can expect that the *Batik* natural dyes would be regenerated by I^-/I_3^- redox couple and allow the electron injection to the semiconductor and complete the electron flow through an external circuit. **Table 2** lists the solar cell parameters of some *Batik* dyes. The order of efficiency of the solar cell corresponds to the ease of electron injection from the dyes into the conduction band of TiO_2. Thermodynamically, the LUMO of MC to the conduction band of TiO_2 is closer than the LUMO of CT and TB (**Figure 12**). Thus, facilitating the electron injection from the dyes to the semiconductor oxide. However, the cell efficiency is still low. It is probably due to the poor cell construction as indicated by the low values for all solar parameters (**Table 2**).

Kay and Gratzel [27] has studied photosensitization of TiO_2 solar cells with chlorophyll derivatives and related natural porphyrins. Mechanism for sensitization has been revealed [28]. Here, spectral sensitization of TiO_2 films with natural chlorophylls extracted from algae is reported. The crude chlorophylls extracts are obtained by methanol extraction of the dried algae. The algae were harvested from Krakal beach, Yogyakarta on September 2007. They were washed with water and air-dried before use. **Figure 13** shows the absorption spectra of some chlorophylls extracted from algae and the corresponding sensitized titania film.

Based on the UV–Vis absorption spectra of the algal methanol extract in **Figure 13**, it appears that the spectra show the two main absorption characters in the visible light region, around 416-422 nm and 660-666 nm. These results are consistent with the results of Kay and Gratzel [27] who have extracted chlorophyll a and b from spinach using methanol as a solvent. The visible light absorption ability of each algal methanol extract can be assessed by determining the solution's light absorption coefficient. In this study, the light absorption coefficient was determined by measuring the uptake of algae methanol extract at different concentrations. Then the absorption coefficient can be determined by applying the Lambert–Beer law (A = *a*.b.c., where A is the absorbance, *a* is the absorption coefficient, b is the thickness of the sample and c is the concentration of the solution). The concentration of algae extract that is not a pure isolated chlorophyll extract is expressed in

Dye Sources		**Active Ingredients**	**J_{SC} ($mA.cm^{-2}$)**	**V_{OC} (V)**	**FF**	**η (%)**	**Ref.**
	Rhododendron	carotenoid	1.61	0.585	0.609	0.57	[20]
	Yellow rose	carotenoid	0.74	0.609	0.571	0.26	[20]
	Tangerine peel	flavonoid	0.74	0.592	0.631	0.28	[20]
	Mangosteen pericarp	anthocyanin	2.69	0.686	0.633	1.17	[20]
	Achiote seed	bixin	1.10	0.57	0.59	0.37	[21]
	Chrysanthemum	xanthophyll	0.09	0.31	0.26	0.01	[22]
	Pomegranate leaf	chlorophyll	2.05	0.56	0.52	0.597	[23]
	Mulberry	anthocyanin	1.89	0.555	0.49	0.548	[23]
	Saffron petal	anthocyanin	2.77	0.36	0.52	0.52	[24]
	*Consolida orientalis*	delphinidin	0.56	0.60	0.53	0.18	[25]
	*Adonis flammea*	astaxanthin	0.40	0.59	0.66	0.16	[25]
	*Salvia sclarea*	eupatilin	0.10	0.37	0.54	0.02	[25]
	Green algae	chlorophyll	0.13	0.41	0.21	0.01	[26]

Dye Sources		Active Ingredients	J_{SC} ($mA.cm^{-2}$)	V_{OC} (V)	FF	η (%)	Ref.
	*Maclura cochinensis* (MC)*	phenolic	0.0064	0.10	0.38	0.0100	this work
	*Ceriops tagal* (CT)*	phenolic	0.0032	0.07	0.21	0.0020	this work
	*Terminalia bellerica* (TB)*	phenolic	0.0064	0.10	0.31	0.0080	this work
	*Sargassum mcclurei* Setchell (SM)⁺	chlorophyll	3×10^{-5}	0.06	0.25	0.0009	this work
	*Hypnea esperi* Bory (HE)⁺	chlorophyll	0.013	0.055	0.31	0.0044	this work

P_{input} = 25.6 mW/cm^2, vapor deposited Au counter electrode.
⁺*P_{input} = 50.0 mW/cm^2, vapor deposited Au counter electrode.*

Table 2.
Photoelectrochemical parameters of DSSC with Batik *and other natural dyes.*

the weight concentration of the extract against the volume of solvent (mg/L), so that *a* is also expressed in mg^{-1} mL cm^{-1}. It appears that the methanol extract of the algae *Sargassum mcclurei* Setchell (SM) has the greatest ability to absorb visible light (*a* = 0.027), while the algae *Hypnea esperi* Bory (HE) has the ability to absorb less light (*a* = 0.006). The value of *a* is characteristic and expresses the intrinsic property of a chemical species to absorb light at a particular wavelength. Based on the electronic spectra of the algae's methanol extract in **Figure 13**, it can be confirmed that chlorophyll *a* is the main component of the algae's extracts. The concentration of chlorophyll *a* (C*a*) can be calculated using the equation $Ca = 12.7 \bullet A_{663} - 2.69 \bullet A_{646}$ [29]. It turns out that the value of *a* is in line with the chlorophyll concentration (C*a*) contained in the algae methanol extract. The C*a* algae SM and HE were 2.59 and 0.96 mg/L, respectively. While, the C*a* of extract HA and AF were 0.35 and 0.98 mg/L, respectively. Algae SM has green leaves, but the others are brown to red. Based on the character of visible light absorption and the C*a*, the algae SM has the best character as a DSSC sensitizer. Two dye extracts of SM and HE were set for I-V measurement. The extract of Hypnea espery Bory is chosen for I-V testing due to its rich spectra absorption from UV to visible region compared to the other two algaes. **Table 2** presents the solar cell parameters as the results from SM and HE solar cells. It is confirmed that SM resulted in better solar cell parameters compared to HE as predicted.

The absorption spectra of four methanol extract dyes of algae as adsorbed on TiO_2 surface, depicted in **Figure 13**, are all relatively broadened forward to both red and blue sides of visible region compared to their respective spectra in methanol solution. These indicate pronounced aggregation occurred as the dyes adsorbed on

TiO_2 surface. However, the absorption pattern of SM is quite different. The electronic absorption of methanol extract of SM exhibits similar pattern to its respective spectra on TiO_2 surfaces. Two main peaks of chlorophyll *a* are still observed. The Soret band experienced hypsochromic shift (blue-shifted), while the Qy band was red-shifted. This indicates that chlorophyll *a* of SM adsorbed on the TiO_2 surface with limited aggregation [30, 31]. Absorption spectrum in the visible region resembles the absorption spectra of chlorophyll *a* in a mixture of methanol or ethanol-water upon completion of transition of monomer into aggregates [32]. A weak shoulder around 445 nm, close to the Soret band is also observed. This may ascribe to the presence of chlorophyllin *b* [33]. Chlorophyllin is chlorophyll derivative in which the cyclopentanone ring is opened as well as the carbonyl of the phytyl ester bond [33]. Compared to the parent chlorophyll, the Qy band of chloropyllin is much weaker than the Soret band. The presence of chlorophyllin affects mainly the intensity of the Qy bands of the crude extracts of algae. In general, the Qy bands of chlorophyll contained in methanol extracts of algae are slightly lower related to the synthetic chlorophyll a presented in the previous reference [33]. The presence of chlorophyllin as observed in **Figure 13** is predicted to facilitate aggregation due to intermolecular bonding induced by the –COOH groups. Efficient photosensitization

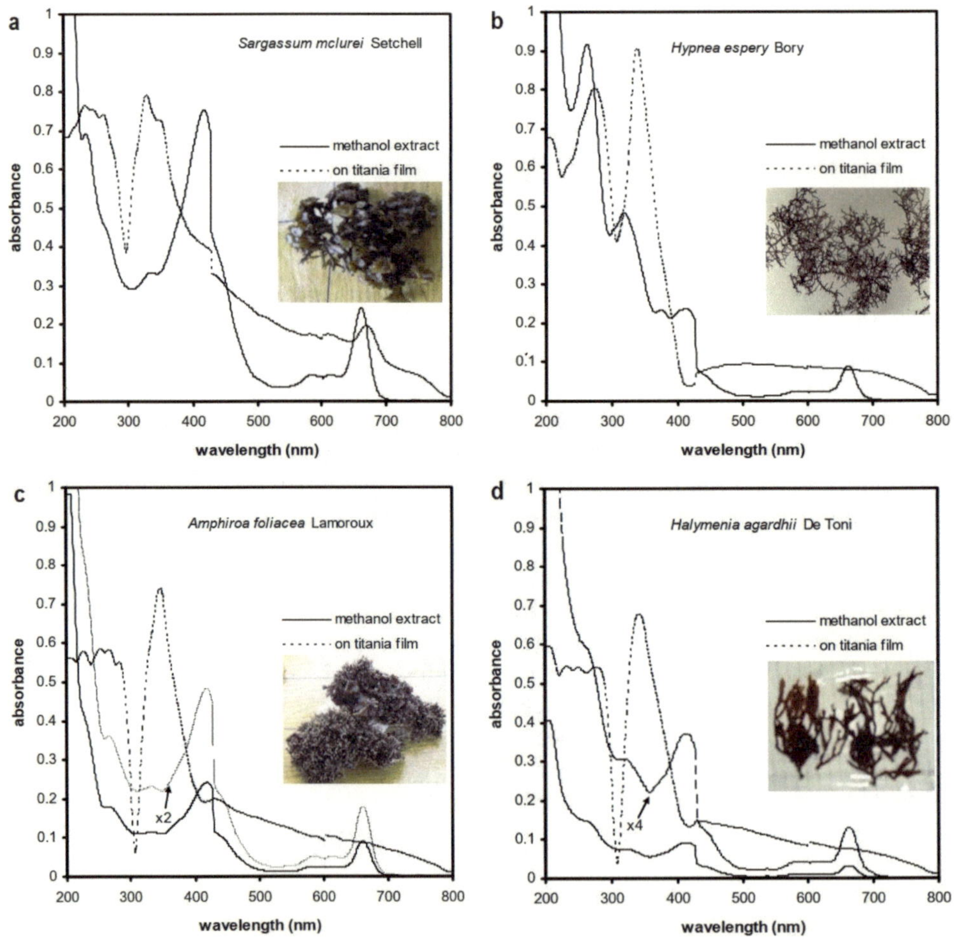

Figure 13.
The electronic absorption of methanol extract of algae adsorbed on titania film and their corresponding solution spectra of: a. Sargassum mclurei *Setchell (SM) b.* Hypnea espery *Bory (HE), c.* Amphiroa foliacea *Lamoroux (AF), and d.* Halymenia agardhii *De Toni (HA).*

may result from efficient electron injection through the bonding formed between TiO_2 and the pigment. Large difference in the photocurrent density of the SM and HE cells rather than in photovoltage suggesting that the solar cell performance of the cells are influenced by the efficiency of the electron injection from the sensitisers into TiO_2 [34].

Natural dyes can be used as a sensitizer, which will require making and purifying dyes more efficiently and rapidly in order to lower production costs, reduce the risk of solar cell toxicity, and use an environmentally sustainable manufacturing method. Some also can be extracted from fruit waste [20, 21], thus it is green technology. Improved efficiency are still intensively researched by employing a cocktail of dyes [20–26, 35], adsorbed dyes on clay [36], optimizing solvent extraction [37]. Amongst, combination of dyes has shown two to three times increased efficiency, while the use of clay has decreased the cell efficiency. It has been shown that TiO_2 is still superior compared to ZnO semiconductor as the photoanode materials [38]. Therefore, discussion will focus on the improvement of natural dyes PEC solar cells due to the use of nanostructured titania. TiO_2 has band gap energy (E_g) in the range of 3.0-3.2 eV. The crystalline phase of TiO_2 found in nature includes anatase, rutile, brookite, and TiO_2-B. Among all the crystalline TiO_2 phases, anatase is the most photoactive crystalline phase. The energy of the upper TiO_2 band gap is 3.2 eV, higher than rutile (3.0 eV). The width of the TiO_2 band gap gives the nature of photostability due to the electron recombination.

The next strategy is to take advantage of the sophistication of nanotechnology, namely utilizing the features of 1D nanostructures such as nanofibers, nanotubes, nanorods; which allows a toll path for electrons from the sensitizer to the back contact of the titania photoanode [39, 40]. The TiO_2 nanorod photoanode gave a value of Voc 0.802 V, Isc 7.01 mA and efficiency of 2.9% [41], whereas TiO_2 nanowire produced Voc 0.752 V, Isc 3.73 mA and efficiency of 1.81% [42]. The TiO_2 nanotube photoanode gave characteristic values of Voc 0.846 V, Isc ~9.63 mA and efficiency of 4.03% [43]. Bijarbooneh et al. [44] used mesoporous TiO_2 nanofibers and obtained Voc 0.76 V, Isc 15.23 mA and were able to increase energy efficiency from 7.28% to 8.14%. The cell performance of nanotubes titania was three-times higher than that constructed from nanoparticle titania (P25) using mangosteen pericarp ethanol extract as the sensitizer [45]. These studies encourage the use of 1D nanostructured TiO_2 to improve performance of the natural dyes solar cells.

Another possibility to improve the natural dyes solar cells is the invention of perovskite material for hybrid DSSCs. Methylammonium lead (II) iodide ($MAPbI_3$) is a perovskite material where the A cation is the organic $CH_3NH_3^+$ cation, B is the Pb^{2+} metal cation, and X is the halide anion such as I^-. Since having a band gap energy of 1.55 eV, which is equivalent to the absorption at a wavelength of 800 nm, this material is potential as visible light absorber. The solar cells efficiency with perovskite structure has achieved up to 25.2% [46]. This high conversion efficiency provides the opportunity to be combined with natural dye sensitised solar cells. Dey et al. [47] has shown that a perovskite and carotene dye layers resulted in a conversion efficiency of up to 5.01%, which was almost ten times than that of solar cells using carotene alone [20]. This new perovskite material feature is expected to be another way to the revival of natural dyes as solar cell sensitisers. However, the presence toxic elements of lead in the perovskite could be a challenge for sustainability.

Recent computational study has shown potential of nanohybrid of graphene quantum dots (GQD), a one type of carbon dots, with porphyrin as the solar cell [48]. It was found that the electron transfer from porphyrin to GQD is faster for larger size of GQD. Nanocomposite carbon dots-polymer [49] has also resulted promising results for quasi solid state solar cells. The carbon dot in the electrolyte composition resulted in improved efficiency up to 6.05% by absorbing unused higher energy of visible light. These findings pave a way to more efficient green natural dyes solar cells.

Our fast-moving time demands creating and innovating science and technology in natural dye's application. Intrinsic properties of the natural dyes of having rich antioxidant are rendering the potential for multifunctional antibacterial textiles. The soft and shady colour of natural dye dyed fabrics with low impact on the environment also drive the fashion industry into the more sophisticated functions of sustainable fashion. It is not only for textile colouring but also for bringing prestige and dignity. The more sophisticated natural dyes function as photosensitisers for photodynamic therapy (PDT) requires intensive purification [50, 51]. Advanced nanotechnology may direct the applications to the photochromic and sensor materials [52, 53].

5. Concluding remarks

Some works on the use of natural dyes for textiles have been presented. The use of natural dyes supports the shifting paradigm in the world fashion to the sustainable fashion. Although, past researches have endorsed essential growth in the application of the natural dyes for fabrics, but still there are a number of technical challenges of natural dye application that must be overcome. The composite formation with green resources such as chitosan, silica may result in enhance dyeing performance to cotton fabrics. Functional such as hydrophobic surface may also be introduced by using natural ingredients such as *Sapindus rarak*.

This work also presents the investigation of the absorption and electrochemical properties of four *Batik* natural dyes to be considered as environmentally friendly photosensitisers for dye-sensitised solar cells. All *Batik* natural dyes extract exhibit absorption peaks in the visible wavelength ensuring their sunlight harvesting ability and HOMO-LUMO energy levels ideal for DSSC. It is noteworthy to blend all the *Batik* dyes to obtain superposition of absorption spectra covering a visible light region from 350 to 800 nm, thus resulting in more efficient panchromatic dyes as required for DSSC. Most of the HOMO-LUMO of the *Batik* dyes have satisfied the thermodynamic requirement as a sensitizer to allow electron transport in DSSC.

Natural dye solar cell technology is still promising as an alternative green and renewable energy. Improved efficiency could be sought through the application of 1D nanostructured titania, the hybrid formation with perovskite organic–inorganic hybrid, and graphene quantum dots or carbon dots. Both, the organometallic perovskite halide and the carbon dots can be used as the co-sensitizer for the realisation of the more efficient natural dyes solar cells.

Acknowledgements

The authors acknowledged financial supports from The Ministry of Research, Technology, and Higher Education of the Republic Indonesia (Ristekbrin) through the National Competitive Research Grant (PD 2020), International Foundation for Science (IFS) Sweden, and Universitas Gadjah Mada – Science Techno Cluster & Departmental research grants. Special thanks also to Alfi Fatihah and Stepanus Fredi Manurung for some experimental works on textile.

Author details

Indriana Kartini[1]* and Adhi Dwi Hatmanto[2]

1 Department of Chemistry and Indonesia Natural Dye Institute, Universitas Gadjah Mada, Yogyakarta, Indonesia

2 Department of Chemistry, Universitas Gadjah Mada, Yogyakarta, Indonesia

*Address all correspondence to: indriana@ugm.ac.id

References

[1] Bechtold T, Mahmud-Ali A, Komboonchoo S. Sustainable dyes from agrifood chain co-products. In: Waldron KW, Moates GK, Faulds CB, editors. Total Food: Sustainability of Agri-Food Chain. RSC Publishing, Cambridge: UK; 2009. p. 211-218. DOI: 10.1039/9781849730785-00211

[2] O'Regan B., Gratzel M. A low-cost, high-efficiency solar cell based on dye-sensitized colloidal TiO_2 film. Nature. 1991;353: 737-739. DOI: 10.1038/353737a0

[3] Tennakone K, Kumara GRRA, Kumarasinghe AR, Sirimanne PM, Wijayantha KGU. Journal of Photochemistry and Photobiology A. 1996;94(2-3):217-220. DOI: 10.1016/1010-6030(95)04222-9

[4] Kasmudjiastuti E. Characterization of Tingi (*Ceriops Tagal*) Bark As Vegetable Tanning Material. Journal of Leather, Rubber, and Plastics. 2014;30(2):71-78. DOI: 10.20543/mkkp.v30i2.128

[5] Nazir F. Mangrove studies: A source of tanning material. Reviews the used of mangrove as an ecologically friendly tanning material. Leather International. 2008;June.

[6] Mahltig B, Textor T. Combination of silica sol and dyes on textiles. Journal of Sol-Gel Science and Technology. 2006; 39:111-118. DOI: 10.1007/s10971-006-7744-9

[7] Mahltig B, Böttcher H, Rauch K, Dieckmann U, Nitsche R, Fritz T. Optimized UV Protecting Coatings by Combination of Organic and Inorganic UV Absorbers. Thin Solid Films. 2005; 485: 108-114. DOI: 10.1016/j.tsf.2005.03.056

[8] Kartini I, Ilmi I, Kamariah, Kunarti ES. Wash fastness improvement of malachite green-dyed cotton fabrics coated with nanosol composites of silica–titania. Bulletin of Material Science. 2014;37(6):1419-1426. DOI: 10.1007/s12034-014-0091-5

[9] Kartini I, Halimah SN, Rahayuningsih E. Enhanced wash-fastness of cotton fabric dyed with a composite of chitosan-natural dyes extract of *Ceriops tagal*. In: IOP Conference Series: Materials Science and Engineering. International Conference on Chemical and Material Engineering (ICCME 2020); 6th-7th October 2020; Semarang; Indonesia: 2021;1053: 012022. DOI: 10.1088/1757-899X/1053/1/012022

[10] Ye W, Leung MF, Xin JH, Pei LZ. Novel core-shell particles with poly(n-butyl acrylate) cores and chitosan shells as an antibactreial coating for textiles. Polymer. 2005;46:10538–10543. DOI: 10.1016/j.polymer.2005.08.019

[11] Hayn RA, Owens JR, Boyer SA, McDonald RS, Lee HJ. Preparation of highly hydrophobic and oleophobic textile surfaces using microwave-promoted silane coupling. Journal of Material Science. 2011;46:2503-2509. DOI: 10.1007/s10853-010-5100-5

[12] Prusty A, Gogoi N, Jassal M, Agrawal AK. Synthesis and Characterization of Non-fluorinated Copolymer Emulsions for Hydrophobic Finishing of Cotton Textiles. Indian Journal of Fibre Textile Research. 2010; 35:264–271. DOI: -

[13] Gao L, McCarthy J. The "Lotus Effect": Two Reasons Why Two Length Scales of Topography are Important, Langmuir. 2006;22(7): 2966-2967. DOI: 10.1021/la0532149

[14] Sharma G, Zervaki G, Angaridis P, Vatikioti A, Gupta K, Gayathri T, Nagarjuna P, Singh SP,

Chandrasekharam M, Banthiya A. Stepwise co-sensitization as a useful tool for enhancement of power conversion efficiency of dye-sensitized solar cells: the case of an unsymmetrical porphyrin dyad and a metal-free organic dye. Org. Electron. 2014;15(7):1324–1337. DOI: 10.1016/j.arabjc.2021.103080

[15] Ludin NA, Mahmoud AA-A, Mohamad AB, Kadhum AAH, Sopian K, Karim NSA. Review on the development of natural dye photosensitizer for dye sensitized solar cells. Renewable and Sustainable Energy Reviews. 2014;31: 386–396. DOI: 10.1016/j.rser.2013.12.001

[16] Khan AU, Hassan A, Gilani. Pharmacodynamic evaluation of *Terminalia bellirica* for its anti-hypertensive effect. Journal of Food and Drug Analysis. 2008;16:6-14. DOI: 10.38212/2224-6614.2355

[17] Schlaf R, Schroeder PG, Nelson MW, Parkinson BA, Merritt CD, Crisafulli LA, Murata H, Kafafi ZH. Determination of interface dipole and band bending at the Ag/tris (8 hydroxyquinolinato) gallium organic Schottky contact by ultraviolet photoemission spectroscopy. Surface Science. 2000;450:142-152. DOI: 10.1016/S0039-6028(00)00232-6

[18] Wu TY, Tsao MH, Chen FL, Su SG, Chang CW, Wang HP, Lin YC, Sun IW. Synthesis and characterization of three organic dyes with various donors and rhodamine ring acceptor for using in dye-sensitized solar cells. Journal of The Iranian Chemical Society. 2010;7:707–720.

[19] Jolly D, Pelleja L, Narbey S, Oswald F, Chiron J, Clifford J, Palomares E, Demadrille R. Robust organic dye for dye sensitized solar cells based on iodine/iodide electrolytes combining high efficiency and outstanding stability. Scientific Reports. 2014;4(4033):1–7.

[20] Zhou H, Wu L, Gao Y, Ma T. Dye-sensitized solar cells using 20 natural dyes as sensitizers. Journal of Photochemistry and Photobiology. 2011; 219:188-194. DOI: 10.1016/j.jphotochem.2011.02.008

[21] Gomez-Ortiz NM, Vazquez-Maldonado IA, Perez-Espadas AR, Mena-Rejon GJ, Azamar-Barrios JA, Oskam G. Dye-sensitized solar cells with natural dyes extracted from achiote seeds. Solar Energy Materials and Solar Cells. 2010;94:40-44. DOI: 10.1016/j.solmat.2009.05.013

[22] Kartini I, Dwitasari L, Wahyuningsih TD, Chotimah. The Sensitization of Xanthophylls-Chlorophyllin Mixtures on Titania Solar Cells. International Journal of Science and Engineering. 2015;8(2):109-114. DOI: 10.12777/ijse.8.2.109-114

[23] Chang H, Lo YJ. Pomegranate leaves and mulberry fruit as natural sensitizers for dye-sensitized solar cells. Solar Energy. 2010;84:1833-1837. DOI: 10.1016/j.solener.2010.07.009

[24] Hosseinnezhad M, Rouhani S, Gharanjig K. Extraction and application of natural pigments for fabrication of green dye-sensitized solar cells. Opto-Electronics Review. 2018;26:165-171. DOI: 10.1016/j.opelre.2018.04.004

[25] Hamadanian M, Safaei-Ghomi J, Hosseinpour M, Masoomi R, Jabbari V. Uses of new natural dye photosensitizers in fabrication of high potential dye-sensitized solar cells (DSSCs). Materials Science in Semiconductor Processing. 2014;27: 733-739. DOI: 10.1016/j.mssp.2014.08.017

[26] Taya SA, El-Agez TM, El-Ghamri HS, Abdel-Latif MS. Dye-sensitized solar cells using fresh and dried natural dyes. International Journal of Materials Science and Applications. 2013;2:37-42. DOI: 10.11648/j.ijmsa.20130202.11

[27] Kay A, Gratzel M. Artificial Photosynthesis. I. Photosensitization of TiO_2 Solar Cells with Chlorophyll Derivatives and Related Natural Porphyrins. Journal of Physical Chemistry. 1993;97:6272-6277. DOI: 10.1021/j100125a029

[28] Kay A, Humphry-Baker R, Gratzel M. Investigation on the mechanism of photosensitization of nanocrystalline TiO_2 solar cells by chlorophyll derivatives. Journal of Physical Chemistry. 1994;98:952-959. DOI: 10.1021/j100054a035

[29] Harborne JB. Phytochemical Methods. 1st ed. London; Chapman and Hall; 1973. 277 p. DOI: 10.1007/978-94-009-5921-7

[30] Ehret A, Stuhl L, Spitler MT. Spectral Sensitization of TiO_2 Nanocrystalline Electrodes with Aggregated Cyanine Dyes. Journal of Physical Chemistry B. 2001;105: 9960-9965. DOI: 10.1021/jp011952+

[31] Sayama K, Tsukagoshi S, Mori T, Hara K, Ohga Y, Shinpou A, Abe Y, Suga S, Arakawa H. Efficient sensitization of nanocrystalline TiO_2 films with cyanine and merocyanine organic dyes. Solar Energy Materials and Solar Cells. 2003;80:47-71. DOI: 10.1016/S0927-0248(03)00113-2

[32] Vladkova R. Chlorophyll a Self-assembly in Polar Solvent–Water Mixtures. Photochemistry and Photobiology. 2000;71(1):71-83. DOI: 10.1562/0031-8655(2000)071<0071:casaip>2.0.co;2

[33] Krautler B. Breakdown of Chlorophyll in Higher Plants—Phyllobilins as Abundant, Yet Hardly Visible Signs of Ripening, Senescence, and Cell Death. Angewandte Chemie International Editions. 2016;55:4882–4901. DOI: 10.1002/anie.201508928

[34] Kamat PV. Photochemistry on nonreactive and reactive (semiconductor) surfaces. Chemical Reviews. 1993;93(1):267-300. DOI: 10.1021/cr00017a013

[35] García-Salinas MJ, Ariza MJ. Optimizing a Simple Natural Dye Production Method for Dye-Sensitized Solar Cells: Examples for Betalain (Bougainvillea and Beetroot Extracts) and Anthocyanin Dyes. Applied Sciences. 2019;9:2515. DOI: 10.3390/app9122515

[36] Saelim N, Magaraphan R, Sreethawong T. TiO_2/modified natural clay semiconductor as a potential electrode for natural dye-sensitized solar cell. Ceramics International. 2011; 37:659-663. DOI: 10.1016/j.ceramint.2010.09.001

[37] Hemmatzadeh R, Mohammadi A. Improving optical absorptivity of natural dyes for fabrication of efficient dye-sensitized solar cells. Journal of Theoretical and Applied Physics. 2013;7: 1-7. DOI: 10.1186/2251-7235-7-57

[38] Gomez-Ortiz NM, Vazquez-Maldonado IA, Perez-Espadas AR, Mena-Rejon GJ, Azamar-Barrios JA, Oskam G, Dye-sensitized solar cells with natural dyes extracted from achiote seeds. Solar Energy Materials and Solar Cells. 2010;94:40–44. DOI: 10.1016/j.solmat.2009.05.013

[39] Kamat PV, Tvrdy K, Baker DR, Radich JG. Beyond Photovoltaics: Semiconductor Nanoarchitectures for Liquid-Junction Solar Cells. Chemical Reviews. 2010;110:6664–6688. DOI: 10.1021/cr100243p

[40] Kartini I. Progress on Nanomaterials for Photoelectrochemical Solar Cell. In: E3S Web of Conferences. The 4th International Conference on Energy, Environment, Epidemiology and Information System (ICENIS 2019); 7-8 August 2019; Semarang; Indonesia: 2019;125, 14015. DOI: 10.1051/e3sconf/201912501001

[41] Wang J, Jin EM, Park JY, Wang WL, Zhao XG, Gu HB. Increases in solar conversion efficiencies of the ZrO_2 nanofiber-doped TiO_2 photoelectrode for dye-sensitized solar cells. Nanoscale Research Letters. 2012;7:98. DOI: 10.1186/1556-276X-7-98

[42] Wei Z, Yao Y, Huang T, Yu A. Solvothermal growth of well-aligned TiO_2 nanowire arrays for dye-sensitized solar cell: Dependence of morphology and vertical orientation upon substrate pretreatment. International Journal of Electrochemical Science. 2011;6: 1871-1879. DOI:

[43] Flores IC, Freitas JN, Longo C, Paoli MAD, Winnischofer H, Nogueira AF. Dye-sensitized solar cells based on TiO_2 nanotubes and a solid-state electrolyte. Journal of Photochemistry and Photobiology A: Chemistry. 2007;189:153-160. DOI: 10.1016/j.jphotochem.2007.01.023

[44] Bijarbooneh FH, Zhou Y, Sun Z, Heo YU, Malgras V, Kim JH, Dou SX. Structurally stabilized mesoporous TiO_2 nanofibres for efficient dye-sensitized solar cells. APL Materials. 2013;1:1-7. DOI: 10.1063/1.4820425

[45] Kartini I, Evana, Sutarno, Chotimah. Sol-Gel Derived ZnO Nanorod Templated TiO_2 Nanotube Synthesis for Natural Dye Sensitized Solar Cell. Advanced Materials Research. 2014;896: 485-488. DOI: 10.4028/www.scientific.net/AMR.896.485

[46] NREL efficiency chart [internet]. 2019. Available from: https://www.nrel.gov/pv/assets/pdfs/best-research-cell-efficiencies.20190802.pdf [Accessed: 2019-09-25]

[47] Dey A., Dhar A, Roy S, Das BC. Combined Organic-Perovskite Solar Cell Fabrication as conventional Energy substitute. Materials Today: Proceedings. 2017;4:12651-12656. DOI: 10.1016/j.matpr.2017.10.077

[48] Mandal B, Sarkar S, Sarkar P. Theoretical Studies on Understanding the Feasibility of Porphyrin-Sensitized Graphene Quantum Dot Solar Cell. Journal of Physical Chemistry C. 2015; 119:6,3400–3407. DOI: 10.1021/jp511375a

[49] Mohan K, Bora A, Dolui SK. Efficient way of enhancing the efficiency of a quasi-solid-state dye-sensitized solar cell by harvesting the unused higher energy visible light using carbon dots. ACS Sustainable Chemistry and Engineering. 2018;6:10914-10922. DOI: 10.1021/acssuschemeng.8b02244

[50] Ivashchenko O, Przysiecka L, Peplińska B, Flak D, Coy E, Jarek M, Zalewski T, Musiał A, Jurga S. Organic–Inorganic Hybrid Nanoparticles Synthesized with *Hypericum perforatum* Extract: Potential Agents for Photodynamic Therapy at Ultra-low Power Light. ACS Sustainable Chemistry and Engineering. 2021;9(4): 1625-1645. DOI: 10.1021/acssuschemeng.0c07036

[51] Ormond AB, Freeman HS. Dye Sensitizers for Photodynamic Therapy. Materials. 2013;6(30:817-840. DOI: 10.3390/ma6030817

[52] Kim D-H, Cha J-H, Lim JY, Bae J, Lee W, Yoon KR, Kim C, Jang J-S, Hwang W, Kim I-D. Colorimetric Dye-Loaded Nanofiber Yarn: Eye-Readable and Weavable Gas Sensing Platform. ACS Nano. 2020;14(12):16907-16918. DOI: 10.1021/acsnano.0c05916

[53] Riaz RS, Elsherif M, Moreddu R, Rashid I, Hassan MU, Yetisen AK, Butt, H. Anthocyanin-Functionalized Contact Lens Sensors for Ocular pH Monitoring. ACS Omega. 2019;4(26):21792-21798. DOI: 10.1021/acsomega.9b02638

Chapter 5

Photochromic Dyes for Smart Textiles

Virendra Kumar Gupta

Abstract

Photochromism is a light induced reversible color change phenomenon in photochromic molecule due to light and heat effect and molecular species exist in two forms which have different absorption spectra. The fascinating color change by photochromic molecules in response to specific wavelength of light produces number of applications such as U.V. protective fabrics, ophthalmic photochromic lenses, optical data storing, optical switch, sensors and display. This chapter provides a brief and conclusive review of photochromism their mechanism and application in Textiles. Although photochromic materials are in use since 1960 in lenses and sunglasses, but the development is slow due to technical difficulties and poor commercial application. Now there is renewed interest in photochromic materials which are used in nanofibers in smart textiles and in allied items.

Keywords: photochromic colorants, thermochromic colorants, U.V. radiation

1. Introduction

The wide and increasing application of photochromic and thermochromic colorants in different fields initiates new interest in dyes and pigments. The photochromic compounds got excited when irradiated at particular wavelength in range of 200–400 nm and few compounds in 430–455 nm range. But researchers are more interested to develop organic rather than inorganic photochromic materials because their response in 400–700 nm visible region. The use of photochromic and thermochromic colorants in making smart materials such as medical thermography, photochromic lenses [1], food packaging materials, liquid crystal alignment [2, 3], optical data storage [4, 5], non linear optics [6, 7], photo switching, molecular photonic devices and in photochromic polymers [8] are well known. There is demand for application of photochromic and thermochromic colorants in making smart textiles,which are designed to sense and respond to external environmental conditions and stimuli. Photochromic and thermochromic colorants are prone to change their colors temporarily and reversibly in presence of UV light, visible light, acids [9], alkalis, water, mechanical strain, temperature and in electric field. These dyes became colored when exposed to these environmental conditions temporarily and revert back upon disappearance of external environment. The photochromic dyes [10] are categorized as inorganic and organic molecules. In inorganic types the important are metal oxide, alkaline earth metals, sulphides, copper compounds and mercury compounds. The organic types are effective and environment friendly and they belong to the families of spiropyrans, spirooxazines, chromomenes, fulgides, fulgimides and diarylethenes. Spiropyrans, spirooxazines and chromomenes are sensitive to thermal effect and reverse to colorless state under heat or visible light however fulgides, fulgimides and diarylethenes are thermally stable. Out of these spiropyrans are having more scientific interest than any other class.

Spiro compounds have pyran ring and linked to another heterocyclic ring through spiro group. Spirooxazines molecules contain nitrogen atom at the place of carbon in spiro group. These molecules (colorless) have non planer structure and that inhibit delocalization of Л electrons in the molecules. In presence of UV light, molecules absorb photon energy and breaking of –C-O- bond in pyran ring takes place and there is formation of colored planar structure molecule. The planarity of molecule allow delocalization of Л electrons and molecule become colored. This is short term phenomenon and after absorption of heat or visible light molecules convert into original structure (colorless) as shown in **Figure 1**.

The photochromism [11, 12] may be defined as a reversible light –induced color change or reversible transformation between two different molecular structures with different absorption spectrum in reversible manner due to electromagnetic radiations. Photochromic materials are kind of chromic materials in which photochromic and thermochromic materials are of paramount importance. In photochromism the colorless molecule became colored in presence of UV Portion of light however in the thermochromic molecules heat is responsible for change of color [4].

The general physicochemical reaction of photochromic molecules are as given in equation no. 1 [13, 14].

$$X(\Lambda_1)\ (\text{colorless})\ \ \underset{\Delta hY_2}{\overset{hy_1}{\rightleftharpoons}}\ \ Y(\Lambda_2)\ (\text{colored}) \qquad (1)$$

Λ_1 and Λ_2 are the wavelength of maximum absorption by corresponding molecules and hY_1 and hY_2 are the energy absorbed by the molecules during transformation. Y_1 is the frequency of wave in U.V. region and Y_2 is the frequency of wave in either U.V. or visible region and Δ is heat requirement. The factors which influence reaction 1 are [15].

a. Wavelength of incident light

b. Speed of recovery or fatigue resistance

c. Long term stability of molecule to produce high number of cycles

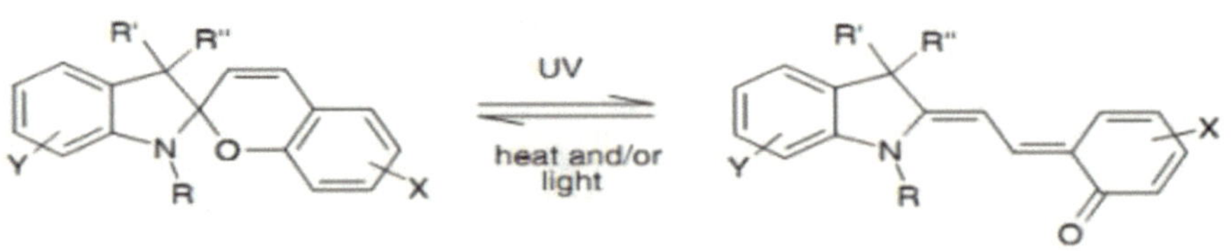

Colorless spiropyran molecule

Colored molecule

R =alkyl group R″ and R″ methyl group

X,Y= H, X, NO_2 etc.

Figure 1.
Conversion of colorless spriopyran molecule into colored molecule.

2. Types of photochromism

2.1 Positive photochromism

In this photochromism photochromic molecule absorb UV light whose λ_{max} falls in UV region and colorless molecule became colored and on reversal during bleaching process in visible wavelength it become colorless.

2.2 Negative photochromism

It is opposite to the positive photochromism instead of coloration discoloration observed on exposure to UV light i.e. the original molecule is colored and after exposure to UV light it loss their color.

2.3 (c) Photo responsible materials

In photochromic reactions, there is conformational modifications in the structure of photochromic molecules and due to that there are change in physico-chemical properties of two form of the photochromic molecules and the change observed in physico- chemical properties of both molecules (colored and colorless) are called photo responsible materials [16].

3. Mechanism of Phtochromism

Photochromic reaction leads to change in electronic absorption spectra of molecules. The formation of new absorption band due to transition of electrons from various vibrational levels in the excitation of colorless molecules from S_1A level to excited state S_1B after absorption of energy photon in UV region [17] and then after the colored molecules deactivated to ground state S_0B. Subsequently there is spontaneous energy release process and molecules come to original S_0A ground state. Thermodynamically the molecules in B state due to higher energy are less stable and after releasing energy became colorless and more stable. The transition from state B to A takes place via a transition state X, whose energy is higher than the triplet state of colored form S_0B and it is thermally activated. There are six mechanism which responsible for the photochromic effect and they are,

- Triplet triplet photochromism.
- Hetrolytic cleavage
- Hemolytic cleavage
- Trans-cis isomerismation
- Tautomerism
- Photodimerisation

The transition of molecules after absorption of light in different energy levels is shown in **Figure 2**.

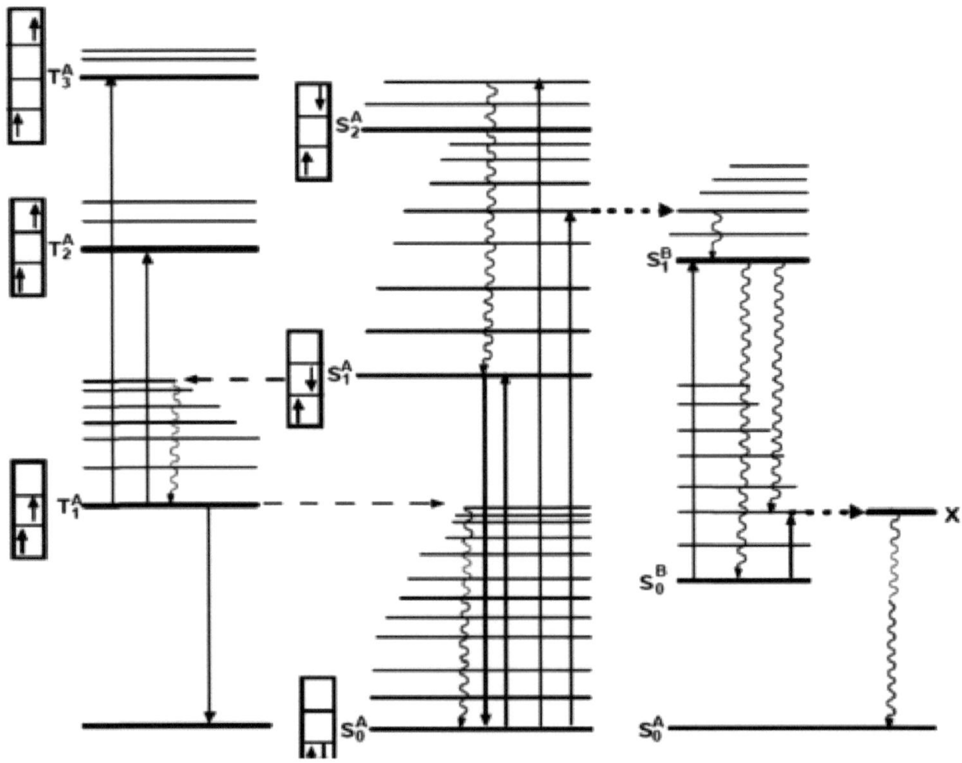

Figure 2.
Representations of electronic, radiative and non radiative transitions in photochromic materials.

4. Classification of photochromic materials

Classification of photochromic materials are based on back reaction i.e., if it from colored state to colorless state is brought in the presence of light then it is called P type photochromic materials (**Figure 1**), where as if back reaction occurred due to heat energy, it is called T type of photochromic materials.

The p type photochromic materials exist in two reversible forms upon irradiation and have good thermal stability with better fatigue resistance. In p type photochromic materials [18, 19] during exposure to UV light ring opening takes place. After ring opening, molecules absorb visible wavelength light and became colored. This state of molecule is temporary and again it became closed ring system (colorless). Most of the p type photochromic materials do not follow trans- cis isomerism and which follow they have they have open ring structure in colorless form after exposure to UV light they became closed ring structure (colored). The reverse reaction i.e. ring opening is promoted by visible wavelength [20, 21].

In T type photochromic materials photochromic reactions takes place due to thermal irradiation or by photo irradiation with visible light [22, 23]. It shows reversible equilibrium between trans and cis isomerism of different stability.
In T type photochromism there is no breaking of bonds occur however there is rearrangement of electrons between energy levels and alteration in geometric arrangement of the molecules. The thermal reversal of molecule takes place in dark. The important T type photochromic materials are perimidinespirocyclohexadienes, spirodihydroindolizines and anils. There are some requirement in T type photochromic materials which are,

- Smaller concentration should produce intense color with effective color change.
- The conversion from colorless to colored form upon exposure to U.V. light must be very quick.
- The photochromic molecules should have minimum time to loss their absorbance half i.e., photo chromic effect should be lost as soon as removal of activating light takes place.
- The photochromic molecules should have quick response in presence of U.V. radiation.
- The life time of photochromic molecules should be longer (Approx. 2 years) in both colored and colorless form. It should also have good resistance to fatigue.
- The photochromic molecules should be less vulnerable to variation in temperature.

5. Types of photochromic materials (T types)

5.1 Spiropyrans

It is first T type photochromic molecules and widely used in the industry. It have application in memory disks, optical switches, sensors and as a photochromic dyes in textiles and plastics. The general structure contain a second ring structure, which attach to pyron core at position two [24–29]. The coloring of the molecules take place under exposure to U.V. light. The ring opening in molecular structure occur under U.V.light and merocyanine form is created and that exist as cic-cis/trans-trans mixture. The absorption of either U.V. or visible wavelength takes place after ring opening and after absorption, new wavelength in visible region is produced. It has low thermal bleaching rate and at high temperature the color became weaker. In **Figure 3** the ring opening of spiropyrans and formation of conjugated structure is shown.

5.2 Spirooxazines

The chemical structure is similar to spiropyrans, the difference exit that instead of pyron core, it has oxazine group [30, 31] (**Figure 4**). The spirooxazines has very good resistance to photo degradation. The spirooxazines produces fast fading blue

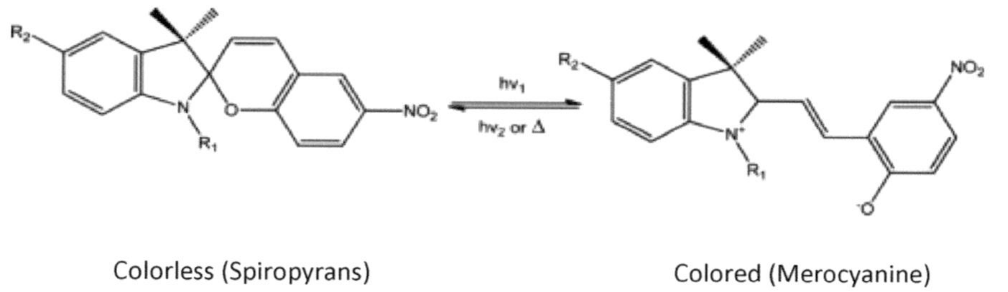

Figure 3.
Photochromic reaction of spiropyrans (from closed ring structure to open ring structure).

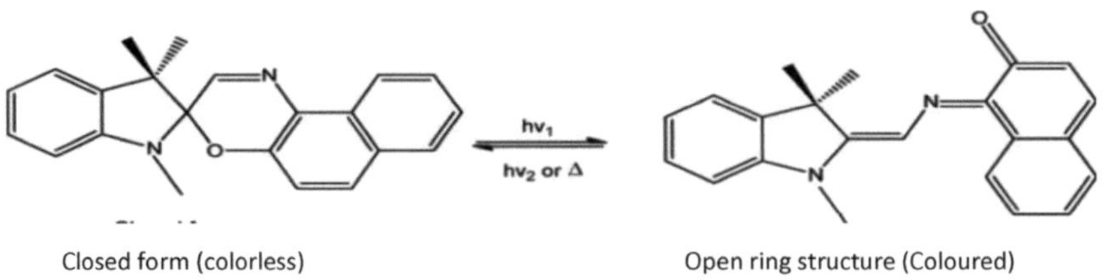

Figure 4.
Photochromic reaction of spirooxazines (from closed ring structure to open ring structure).

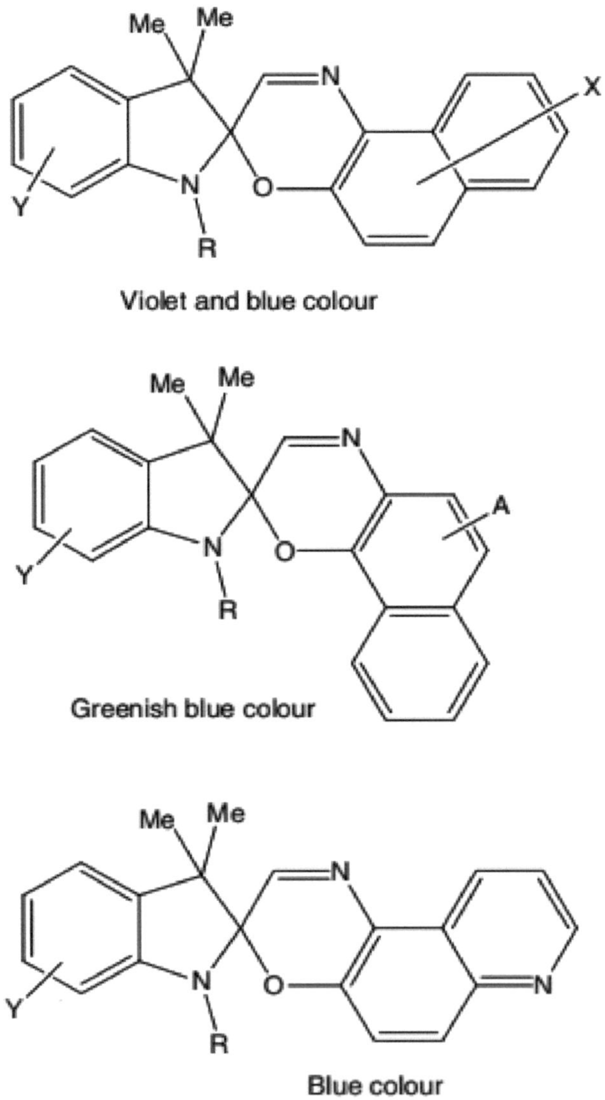

Figure 5.
Different structures of photochromic oxazine (R = alkyl group, X = H, amino, hetaryl Y = H, halogen, a = electron acceptor.

photo coloration. The different chemical structure of spirooxazines are shown in **Figure 5**. The presence of different alkyl groups R at nitrogen atom decide the fading and color strength of molecules [32]. In 1990s, plastic photochromic

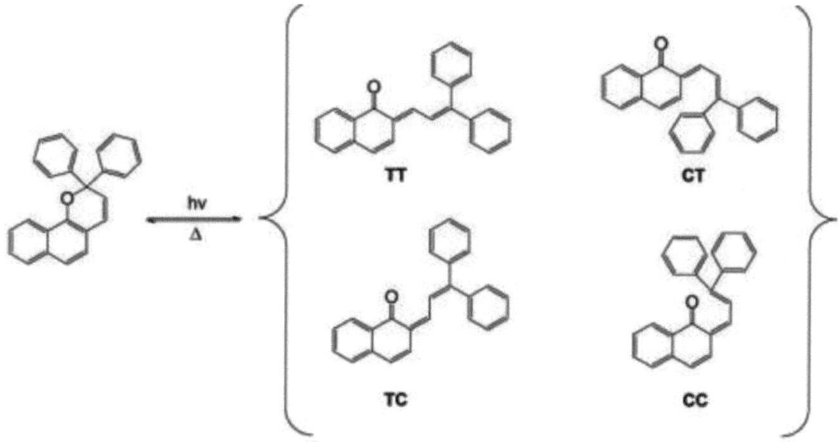

Figure 6.
Photochromic reactions of napthopyrans (TT trans-trans, CT cis-trans, TC trans-cis, CC cis-cis).

ophthalmic lenses were manufactured using spirooxazines. Other applications are photochromic inks, dyes and various cosmetics items.

5.3 Napthopyrans/benzochromenes

It has wide commercial applications such as in plastic photochromic lenses. The photochromism mechanism of benzo and naphthopyron are similar to spiropyrans [33]. They all have breaking of C—O bond in the pyron core. The photochromic reactions of napthopyrans/benzochromenes by the light induced ring opening is shown in **Figure 6**. The ring opening of molecular structure produces more planer structure with greater conjugation of bonds. The planer conjugated structure are capable to absorb visible region wavelength and produces color. The naphthopyrans are less sensitive to temperature.

6. P type photochromic materials

The p type photochromic materials have applications in computing, optical circuitry, memory technology and in ultra high density storage devices. The p type photochromic materials are of two types such as fulgides and diarylethenes. In p type photochromic molecules, the open ring structures are colorless and closed ring structures are colored [34–36].

6.1 Fulgides

The fulgides and fulgimide family belong to P type photochromic materials (**Figure 7**). In 1905 Stobbe synthesized some photochromic fulgides named as phenyl substituted bismethylene succinic anhdides. Heller et al. [37] developed a compound succinimide called fulgimide. It exhibit good photochromic properties. It has shows absorption spectrum of both forms with efficient photoreaction, thermal and photochemical stability. The application of fulgides are In optical switches, sensors, dye inks and memory disks.

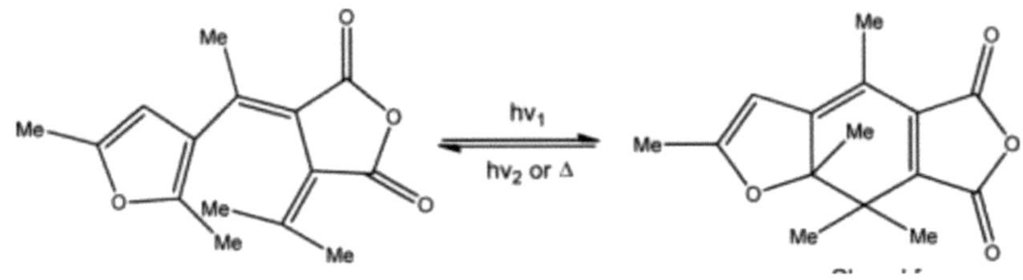

Colorless (open form)

Colored (close form)

Figure 7.
Photochromic reaction of fulgides.

6.2 Diarylethenes

It has hetrocyclic five membered rings such as thiophene or benzothiophene rings and undergo thermally irreversible and fatigue resistant photochromic reactions as shown in **Figure 8**. The thermal stability of both isomers of diarylethens are due to presence of aryl groups and when aryl groups are furan or thiophene, the closed form is thermally stable. However diarylethenes exhibit thermally reversible reactions when the aryl group is phenyl or indole.

7. Effect of temperature

Surrounding temperature influence the performance of photochromic dyes. The bleaching effect of photochromic dyes are accelerated by the temperature of sample [38–40]. Dulic et al. find that ring opening process of diarylethene is temperature dependent whereas ring closing process shows only slight dependence. Ortica [41] reviewed the effect of temperature on the characteristics of various photochromiv materials such as spirooxazines, chromenes and arylethenes, which are as follows,

- In thermo reversible photochromic materials, thermal bleaching increase with increasing temperature.
- Specific temperature can induce spontaneous coloration in thermochromic materials however decreasing temperature will not help in reducing the complexity of photochemical reactions.

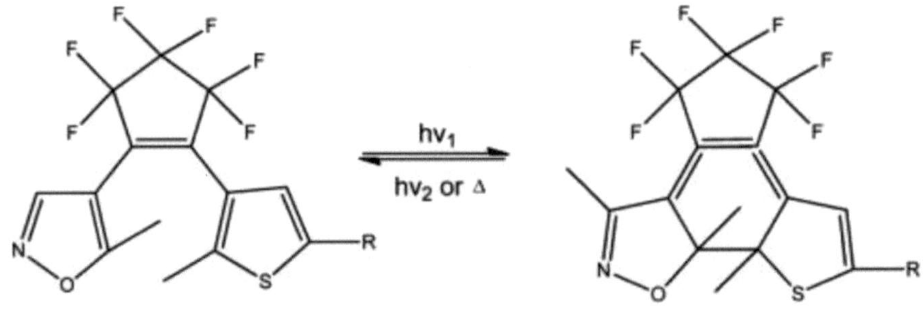

Figure 8.
Photochromic reactions of diarylethenes.

- In nitro substituted chromenes, it was observed that temperature variations influenced the coloration which developed due to substitution in the molecular structure.

- A combination of photochromic and thermochromic materials possess synergistic effect and give superior performance at high temperature.

8. Applications in textiles

8.1 By exhaust dyeing

8.1.1 Method 1

In this method the photochromic dyes are dispersed with dispersing agent [42] and dissolved in water keeping the M:L ratio of dyeing 1:50. The dyebath pH is maintained between 4.5–5.5. The dyebath temperature raised from 40 ^{0} C to 60 °C with 2^{0} C/min gradient and then after it is reduced 1^{0} C/min and final dyeing temperature is maintained at 90 ^{0} C and dyeing is continued at this temperature for 60 min. After completion of dyeing, soaping, rinsing and washing are done to improve fastness properties.

8.1.2 Method 2

In this method the photochromic dyes can be applied as a disperse dye on polyester fabrics by exhaust method of dyeing [43]. The dye is pasted with acetone and then stirred in water with dispersing agent (1%) keeping the M:L ratio of dyeing 1:50.The pH of the dyebath is maintained between 4.5–5.5.The aqueous dyebath is boiled to evaporate acetone, subsequently temperature is raised to 120°C and dyeing is done at this temperature for 45 minute. After dyeing reduction clearing treatment is given at 70 °C for 20 min. and finally sample is soaped, rinsed and washed.

8.2 By continuous dyeing

In this method photochromic dyes are dissolved in acetone and then mixed with binder solution and padded with padding mangle at appropriate pressure. After padding fabric is dried at 80 C and cured at 140 C for 3.0 minute in hot air oven or stenter machine [44].

8.2.1 As a disperse dyes in printing

Photochromic dyes can be used as a disperse dyes which are insoluble in water. The photochromic dyes are disperse with dispersing agent and wetting agent of anionic nature. The dye dispersion is milled on a roller mill by using ceramic balls in glass jar. The dye dispersion is mixed with sodium alginate thickener solution to get printing paste. The polyester or nylon fabric can be printed with printing paste. The fabric is dried at 100 °C and cured at 140 °C for 5 minute. After reduction clearing treatment, printed samples were soaped with nonionic detergent and finally neutralization is done.

8.3 Application of thermochromic materials

Photochromic materials have applications in both textiles and non textiles. In the textile field new fashionable colors in T-shirts, hand bags and caps are produced by dyeing with photochromic dyes. There is application of photo colorable textured yarn in knitting, weaving and embroidery. Polypropylene threads are produced by mass coloration by adding photochromic dyes in melt polymer solution, which on exposure to U.V. light produces different colors. Photochromic colorants are also used in developing camouflage patterns for military protective clothing. The patterns change their colors on exposure to sunlight and match with surrounding environment. Photochromic microencapsulated fabrics are produced which change their color on exposure to sun light. In non textile field photochromic materials are used in ophthalmics, surface coating applications and dye lasers.

8.4 Color measurement

Due to reversible color changing properties of photochromic dyes, it is very difficult to measure color value of the shade produced due to photochromic effect. For measuring the color value, it is essential to control several parameters such as temperature and time interval between U.V. irradiation of sample and measurement. A.F. Little et al. [45] developed a technology to measure the color value of photochromic textiles using independent U.V. irradiation with traditional spectrophotometer. The temperature of sample measuring cabinet was controlled by localized air heating system. The temperature of sample cabinet is maintained at 24 °C and time interval of 30 sec. is kept between irradiation and measurement of sample which can be seen in **Figure 9**.

8.5 Washing fastness test

Due to dynamic color change properties of photochromic dyes, it is difficult to measure the fastness properties. The traditional assessment method of color fastness using gray scale standards [46] are not appropriate, therefore it is measured by comparative test method. In this method we measure the color difference of sample before and after wash and compare with color difference before U.V. irradiation and after 1 min. Exposure to U.V. irradiation. The level of photo coloration

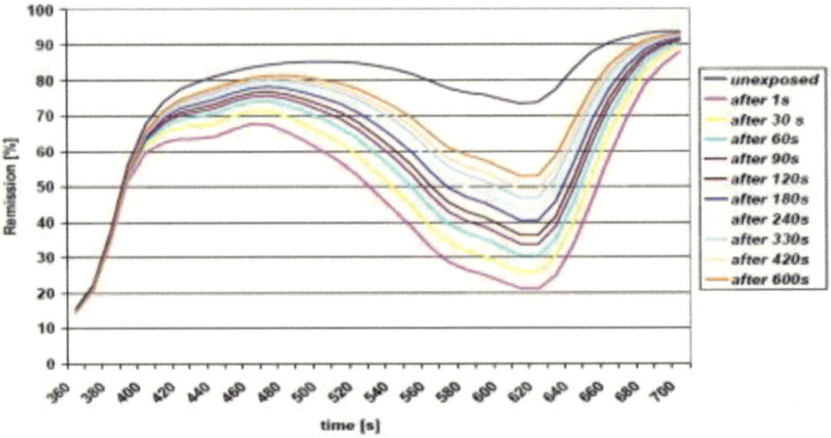

Figure 9.
Color measurement (color bleaching) of photochromic dyes fabric samples in 30 sec. Time interval.

developed by U.V. irradiation varies with photochromic colorant classes. It was revealed that in selected spirooxazine colorants the degree of photo coloration increases with initial washing and subsequently decreases. In case of napthopyrans, the degree of photo coloration decreases continuously with successive number of washings. In case of printing the washing fastness more depends on binder quality.

8.6 Light fastness/photostability

The conventional method of exposing the sample to accelerated fading instruments (Xenotest or MBTF) is not applicable to photochromic colorants. In photochromic colorants due to dynamic color change properties, for light fastness measurement a normalized value of color value to be calculated. The normalized value is defined as the degree of photo coloration after a particular time of exposure on the xenotest instumemt to the fraction of initial degree of photo coloration i.e., $\Delta E/\Delta Eo$.

ΔEo = color difference before and after U.V. Irradiation without exposing to xenotest instrument.

9. Thermochromism

Thermochromism may be defined as the reversible change in the color of compound due to temperature change [47]. The phenomenon of thermochromism may occur even in small temperature interval. The thermochromism can be depicted as shown in **Figure 10**.

The following type of materials can exhibit thermochromism properties [48].

- Organic compounds
- Inorganic compounds
- Polymers
- Sol gels

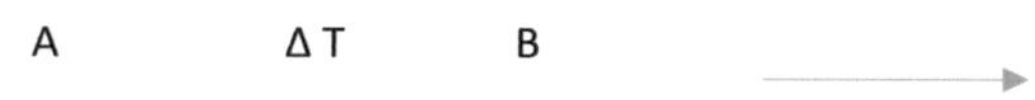

Figure 10.
Reversible color change from species a to B due to thermochromism. Here A = colored or colorless. B = colored or colorless.

10. Organic compounds

It may occur as a result of equilibrium between molecular species such as acid – base, keto-enol and different crystal structure. The organic thermochromic materials have application in fibers, optics and optical sensors. The organic thermochromic compounds show sharp color change due to temperature variation. The different organic thermochromic compounds may be classified as follows.

10.1 Liquid crystals

Some organic materials when pass from crystalline solid to isotropic liquid state, they form stable intermediate phases (mesophase). Transitions between phases are

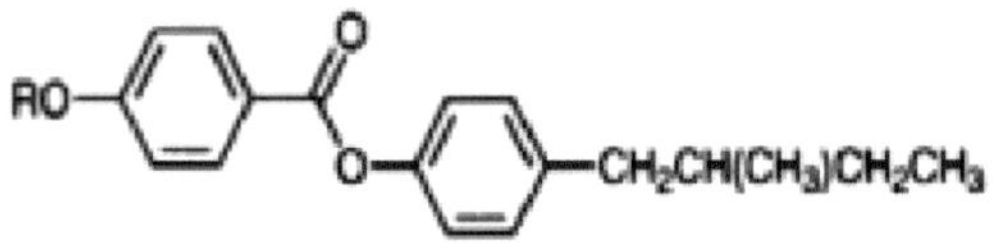

Figure 11.
Chiral molecules in cholesteric mesophases form.

brought either by influence of temperature or solvent respectively [49]. The cholesteric (chiral nematic) are the most important type of liquid crystals for thermochromic systems. Here molecules are arranged in helical form. The reflection of light by liquid crystals are influenced by temperature. The pitch of the helical arrangement of the molecules decides the wavelength of reflected light [50]. The liquid crystals are microencapsulated to get the thermochromic effect. They are applied to the fabric with the help of binder (**Figure 11**).

10.2 Stereoisomerism

Organic molecules which posses stereoisomerism, show thermochromism especially ethylenes. When change in temperature takes place, molecules of these compounds switch between different stereoisomers. Generally the required temperature for thermochromism to takes place is more than 150 C. So they can not be used for textiles application [51].

In another case, the crystal violet lactone is pH dependent rather than temperature dependent. At pH above 4.0 crystal violet lactone is colorless and below 4.0 pH, it is violet.

10.2.1 Rearrangement

Due to molecular rearrangement of organic compounds more conjugated structure resulted and formation of new chromophores take place. Such type of molecular arrangement occur due to temperature variation, change in polarity of the solvent or pH of the solution.

11. Inorganic thermochromic system

Thermochromic behavior are exhibited by solid or liquid inorganic molecules. In such type of molecules thermochromic properties are due to following mechanism [52],

- Phase transition
- Change in ligand geometry
- Equilibrium between different molecular structures
- Change in the number of solvent molecules in the co-ordination sphere.

These compounds show theromochromic properties at high temperature (150^0 C), therefore they are not suitable for textile application.

12. Microencapsulation

In reversible thermochromic compounds a colorless dye precursor and color developer both are dissolved in hydrophobic non volatile organic solvent and resulted solution is encapsulated [53]. On heating, melting of organic solvent occurs and there is appearance of color in thermochromic compound. On cooling solvent solidify and system comes to original color. Microencapsulation has some advantages that it protect sensitive coloring agents from external environment and allows several thermochromic colorants to be combined together and produces several narrow color ranges.

The organic solvents used in microencapsulation are alcohols, hydrocarbons, ester, ketones, thiols and alcohol –acrylonitrile mixture. The important thermochromic colorants are N-acyl leuco-methylene blue derivatives, fluoran dyes and diphenylmethane compounds. There are large number of compounds work as color developers such as phenol derivatives specifically bisphenol A and bisphenol B. In some recent work there are use of 1,2,3 triazoles such as 1,2,3 benzotriazole, dibenzotriazole, thioureas and 4 hydroxy coumarin derivatives.

13. Application of thermochromic pigments on textiles

13.1 By exhaust method

- Cationic agent 5–8% (owf)
- Thermochromic Pigments 10–15% (owf)
- Non ionic leveling and dispersing agent 10–15% (owf)
- Acrylic Binder 10–15%

Thoroughly pretreated fabric taken in water keeping M; L ratio 1:20 and we add cationic agent at temperature of 60 °C and run in the aqueous media for 15 min to get positive charge on fabric. After treatment fresh water is taken in the bath and thermochromic pigments are added, temperature is maintained at 70 °C run fabric for 10–15 min. During dyeing non ionic dispersing agent and leveling agent are added. Finally acrylic binder is added and we run the fabric at 70 C for 15 min.The fabric sample is soaped and washed. In microencapsulated fabric the melting temperature of solvent control the temperature at which decolonization/colorization of thermochromic colorant occur.

13.2 Continuous method (In solution)

- Cationic agent 10–15%
- Thermochromic pigment 50%
- Nonionic dispersing and leveling agent 10–15%
- Acrylic binder 20%

13.3 Application technique

Pad ——————→ Dry ——————→ Cure
At room temperature (expression 70%) 80 C (3 min.) 140 C (2–3 min.)

13.4 Printing recipe

a. Thermochromic Colorant 20 part

b. Emulsion thickner 76 part

c. Acrylic binder 4 part

14. Thermochromic cellulose fiber

Marcin Rubacha et al. [54] developed a method to get thermochromic pigment added cellulosic fiber called Lyocell. 1–10% chromicolor AQ-INK magenta type 27 pigmentation was used as a thermochromic modifier.

15. Photochromic polymers

A photo chromic polymer has photo chromic, chromophoric groups inside the polymer backbone chain. The chromophoric group respond to external radiative stimuli during the photo-irradiation of the polymers, there are change in physical chemical and optical properties of polymer in reversible manner. In 1967 Lovrien attempted to produce a polymer chain with photo irradiation sensitive properties by incorporating azo chrysophenine into polymethacrylaic acid. In 1970 Agolini and Gay [55] investigated photo chromic polyamides incorporating azo benzene. The potential application of photo chromic polymers are in photo chromic glasses, UV sensors, halographic recording media, non linear optics and memory devices.

16. Textile printing

Feczko et al. [56] printed cotton fabric by using photo chromic dyes based on Ethyl cellulose –spirooxazine nano particles with light absorbers. They use micro-encapsulation technique to incorporate photochromic colorants. Vikova [57] applied photo chromic pigments on different fabrics such as cotton, polyester and poly acrylonitrile by screen printing using pigment printing method.

17. Sol –gel coating methods

A sol is dispersion of solid particles in liquid where particles are sufficiently small (0.1–1.0 μm). Due to inorganic nature [58] of layers formed by the sol gel process it possess strong wear resistance and very thin nanometric sized layers. The preparatory materials in preparing sols are inorganic metal salts or organometallic compounds. The preparatory materials are submitted to series of hydrolysis and polymerization reactions to produce colloidal suspension or sol, once the polymerization is completed the colloidal form of the sol developed. Cheng et al. [59–61] prepared silica as a matrix material for fixing the photo chromic dye 5 Chloro 1,3 dihydro-1,3,3 trimethyl spiro on the surface of wool fabric through sol–gel process.

18. Mass coloration

Mass coloration or dope dyeing is method in which colorants are added in the spinning composition before extrusion of filaments. Photo chromic polypropylene [62] thread was prepared by this method. Vikova et al. [63, 64] prepared polypropylene multi filaments by adding photo chromic pigments during dope preparation. Concentration of photo chromic pigments was 0.25,0.5,1.5 and 2.5% by weight.

19. Conclusion

Photochromic and thermochromic materials presenting a new field of research which are yet not fully explored in field of textiles and other allied fields in context of application and durability. The photochromic and thermochromic colorant occupied a niche position in colorant industry. Their application mainly concerned with fashion, leisure and sports garments. Incorporation of photochromic colorants in nanofiber based photochromic textiles can be made use of in smart textiles because of their ability to react with external stimuli, which may work as chromic sensor. Efforts can be done to improve properties such as light fastness and simplistic application procedure. Presently two types of thermochromic system liquid crystals and molecular rearrangement types have successful commercial applications. Japan and U.S, A had developed wide range of thermochromic and photochromic materials, but innovations are required for commercial applications. The factors influencing the equilibrium between colored and colorless form of colorant has to be further explored. Presently only small number of commercial organizations are engaged in research and development work of improved compositions. There is need of study with respect to formulations, encapsulations and application of photochromic and thermochromic materials in textiles.

Author details

Virendra Kumar Gupta
Department of Textile Chemistry, M.L.V. Textile and Engineering College, Bhilwara, Rajasthan, India

*Address all correspondence to: virendra1970@rediffmail.com

References

[1] Eds H.R. Mattila, Intelligent Textiles and Clothing, Woodhead Publication Limited, Cambridge, England. 2006, 296.

[2] Pardo R., Zayat M. and Levy D., Chem. Soc. Rev. 40(2011)672.

[3] Lee E.M., Gwon S.Y., Ji B.C.,Wany S. and Kim S.H., Dyes Pigments 92(2012) 542.

[4] Janus K.,Sworakowski J. and Luboch E., Chem. Phy., 285 (2002)47.

[5] Dong H., Zhu H., Mang Q., Gong X and Hu W., Chem. Soc. Rev., 41 (2012) 1754.

[6] Sun Z., Li H., Liu G., Fan C. and Pu. S., Dyes Pigments 106 (2014)94.

[7] Nigel Corns S., Partington S.M. and Towns A.D., Coloration Technology 125 (2009)249.

[8] Cheng T., Lin T., Brady R. and Wang X., Fibers Polymers 9 (2008)301.

[9] Eds J.C. Crano and R.J. Guglielmetti, Organic Photochromic and thermochromic compounds, Volume 1, Main photochromic families, Kluwer academic press. New Yark, N.Y.1999.

[10] Periyasamy A.P., Vikova M. and Vik M., Textile Progress, 49 (2017)54.

[11] Corns S.N., Partington S.M., Towns A. D., Color Technol, 2009, 125, 249.

[12] Rawat M.S.M., Mal S. and P. ,Open Chem.J.2 (2015) 7.

[13] Corns S. Nigel, Partington S.M. and Towns A.D., Coloration Technology,125 (2009)249.

[14] Hadjoudis E. and Mavridis I.M., Chem. Soc. Rev., 33(2004)579.

[15] Christie R.M., Color Chemistry, The Royal Society of Chemistry, London, 2001.

[16] Durr H., General Introduction in Photochromism: Molecules and System, B.L. Henri and H. Durr, Eds., Elsevier, B.V. Amsterdam, 2003,1.

[17] Vikova M., Photochromic Textiles, Heriat Watt University, Scottish Burdens, Edinburgh, 2011.

[18] Evans R.C., Douglas P. and Burrows H.D. (eds), Applied Photochemistry, Springer International Publishing Work, New York, NY, 2013.

[19] Zhang J.Z., Schwartz B.J., King J.C and Harris C.B., J. American Chemical Society.114 (1992)10921.

[20] Tamai N. and Miyasaka H., Chemical Rev., 100(2000)1875.

[21] Preigh M.J., Stauffer M.T., Lin F.T. and Weber S.G., J. Chem. Soc., Faraday Trrans, 92(1996)3991.

[22] Adamo C. and Jacuemin D., Chem. Soc.Rev., 42(2013)845.

[23] Zhang C., Zhang Z., Fanand M., Yan W., Dyes Pigments, 76(2008)832.

[24] Christie R.M., Advances in dyes and colorants, In advances in the dyeing and finishing of technical textiles, M.L. Gulrajani, ed, Woodhead Publishing Limited, Cambridge, U.K., (2013) 1.

[25] Dawson T.L., Coloration Technology, 126 (2010)177.

[26] Oda H., J. Soc. Dye. Color, 114 (1998) 363.

[27] Ohnishi Y., Yoshimoto S. and Kmura K., J. Photochem. Photobiol. A. Chem., 141(2001)57.

[28] X. Li, Ji. Li, Wang Y., Matsuura T. and Meng J., J. Photochem. Photobiol. A. Chem., 161(2004)201.

[29] Chowdhury M.A., Joshi M. and Butola B.S., J. Eng. Fibres Fabrics, 9 (2014)107.

[30] Corns S. Nigel, Partigton S.M. and Jowns A.D., Coloration Technology, 125 (2009) 249.

[31] Billah S.M.R., Christie R.M. and Shamey R., Coloration Technology, 124 (2008)223.

[32] Becker R.S. and Michl J., J. Am. Chem. Soc., 88 (1966) 5931.

[33] Oliveira M.M., Salvador M.M., Vermeersch G., Micheau J.C., Coelho P. J. and Carvalho M., J. Photochem. Photobiol. A. Chem., 198(2008)242.

[34] Ortica F., Dyes and Pigments, 92 (2012)807.

[35] Little A.F. and Christie R.H., Coloration Technology, 126 (2010)164.

[36] Eds Crano J.C. and Guglielmetti R.J., Organic Photochromic and thermochromic compounds, Volume 2, Physico chemical studies, biological applications and thermochromism, first Roc 2 Kluwer academic press. New Yark, N.Y. 1999.

[37] Hart R.J., Heller H.G. and Salisbury K., Chem. Commune. (London) (1968)1627.

[38] Irie M. and Mohri M, J. Org. Chem. 53(1988)803.

[39] Padwa A., A.Au, Lee G.A and Owens W. J. Org. Chem. 40(1975)1142.

[40] Tian H. and Yang S., Chem. Soc. Rev. 33(2004)85.

[41] Ortica F., Dyes Pigments 92(2012) 807.

[42] Little A.F. and Christie R.M., Coloration Technology, 126(2010)164.

[43] Aldib M. and Christie R.M., Coloration Technology, 127(2011)282.

[44] Little A.F. and Christie R.M., Coloration Technology, 126(2010)157.

[45] Little A.F. and Christie R.M., Coloration Technology, 127(2011)275.

[46] Standard methods for determination of color fastness of textiles and leather 5th edition, Bradford, SDC, 1990.

[47] Day J.H., Chem. Rev. 63(1963)65.

[48] Aitken D., Burkinshaw S.M., Griffiths J. and Towns A.D., Rev. Prog. Coloration, 26(1996)1.

[49] Seeboth A., Luckowska A.K., Ruhmann R., Lotzch Chin D., J. Polymer Science, 25 (2007)123.

[50] Mather R.R., Rev. Prog. Coloration, 31 (2001) 36.

[51] Chowdhary M.A., Joshi M. and Butola B.S., Journal of Engineered Fibres and Fabrics 9(2014)120.

[52] Day J.H., Chem. Rev., 68(1968) 649.

[53] Vesel A. and Gunde M.K., Dyes and Pigments, 86(2010) 271.

[54] Rubacha M., Polymer for advance technology, 18 (2007)323.

[55] Agolini F. and Gay F.P., Macromolecules 3(1970)349.

[56] Feczko T., Samu K., Wenzel K., Neral B. and Vocin B., Coloration Technology 129(2013)18.

[57] Vikova M., Photochromic Textiles, Heriot Watt University, Scottish Bordens Campus ,Edinburgh 2011.

[58] Attia S., Wang J., Wu G., Shen J. and Ma.J., J. Matter. Sc. Technol. 18(2002) 211.

[59] Cheng T., Lin T.,Brady R. and Wang X., Fibers Polym. 9(2008)301.

[60] Cheng T., Lin T., Fang J. and Brady R., Text. Res. J. 77(2007)923.

[61] Cheng T., Lin T., Brady R. and Wang X., Fibers Polym. 9(2008)521.

[62] A Glossary of AATCC Standard Terminology, Resaerch Triangle Park, NC,2007.

[63] Periyasamy A.P., Vikova M. and Vik. M., 24 th International Fedration of Association of Textile Chemist and Colourist Congress, Radim Hardina, ed, University of Pardubice, Czech Republic (2016)1.

[64] Vikova M., Periyasamy A.P., Vik M. and Ujhelyiova A., J. Text. Inst. (2016)1.

Chapter 6

Application of Smart and Functional Dyes in Textiles

Deepti Pargai

Abstract

Our future will be based on functional and AI based smart products, where every industry wants to develop these kinds of products. Textile industry also cannot remain untouched with this technological innovation. Dyes have been utilising for coloration of textiles since ancient time. But at present with various advancement in technology as well as requirements of consumers, the need for functional and smart dyes arises. Various current researches are based on application of smart and functional dyes on textile to develop smart and functional textiles. The dyes which add the functional and smart properties to the textiles can be called as functional and smart dyes. Functional and smart dyes are available in both synthetic and natural form. But with the environment concern, the researchers are going on to find out natural source of these dyes. Functional dyes such as UV protective dye, antimicrobial dye, moth repellent dye offer specific function after application on textiles. Smart dyes like photochromic, thermochromic, electrochromic and solvatochromic etc. are playing very imp role to develop a smart textile which can offer reversible colour change which leads to impart various properties such as thermoregulation, camophlage properties into the textiles. Functional dyes generally limited to the textiles sector but smart dyes are not just restricted to it. Application of smart dyes is extended to various fields such as automobiles, robotics, aircrafts, medicine and surgery etc. This chapter will mainly deal with the types, application methods and application area of functional and smart dyes especially in connection with textiles.

Keywords: functional dyes, smart dyes, photochromic dyes, thermochromic dyes, electrochromic dyes, solvatochromic dyes, UV protective dyes, antimicrobial dyes, smart textiles, functional textiles

1. Introduction

Smart and functional textiles is the need of the future. There are various methods to develop smart and functional textile which start right form the fibre stage and end in the dyeing and finishing stage. Among these methods, application of smart and functional dyes is considered the most affordable method to develop smart and functional textiles. Traditionally, Dyes has been utilising in textiles generally for coloration purpose. Earlier, consumer wants to buy the textiles mainly due to its aesthetic aspect. Presently, consumer has smart choices. A textile product only having aesthetic property could not allure them. In near future, the product should be smart and functional one to attract the consumer. Smart textiles refer, a textile which can act according to their nearby environment while functional textiles refer, a textile which provides a particular function specially protection from nearby environment. Smart and functional textiles are considered as the part technical textiles. Since last few decade the researchers have been searching the functional as well as smart aspect of dyes. The dyes which can offer smart and functional textiles can be considered as smart and functional dyes. Earlier dyes were categorised as natural dyes and synthetic dyes. Natural dyes were categorised as based on origin (vegetable, mineral and animal), colour (yellow, red, blue), chemical constitution, (indigoid, anthraquinone, alphanapthaquinone, dihydropyrans, anthocynidins, cretonoids), application method (direct, basic, acid, vat, disperse, reactive, mordant) while synthetic dyes were classified as water soluble (direct, basic, acid, reactive), water in soluble (vat, disperse, sulphur) and in situ colour formation (azoic dyes) [1, 2]. But Presently and in near future, the dyes can as be categorised as smart dyes and functional dyes on the basis of providing AI based smartness as well as functionality to the textiles. The types of smart and functional dyes are being described in this chapter.

Smart dyes can sense the nearby environment and make changes accordingly [3]. Smart dyes can change the colour due to various external stimuli such as pH, pressure, temperature, solvents, moisture and electricity. These changes can be permanent or temporary on the basis of need [4]. This phenomenon is known as chromism [5]. There are different types of chromism such as photochromism (induced by sunlight or UV rays), thermochromism (induced by changes in temperature), solvato chromism (induced by polarity of the solvent), hygrochromism (induced by moisture), ionochromism (induced by ions), halochromism (induced by pH value), acido chromism (induced by acids), chemochromism (induced by specific chemical agents like dangerous gases, warfare agents, etc.), electrochromism (induced by electricity), piezochromism (induced by pressure), mechanochromism (induced by deformation of substances) [6]. In the field of textiles mainly photochromism, thermochromism, electrochromism and solvatochromism types of chromism is reported [7–10]. Smart dyes are based on these different kinds of chromism. Several natural and synthetic smart dyes have been discovered. On the basis of external stimuli smart dyes are classified as photochromic dye, thermochromic dye, solvato chromic dye, hygro chromic dye, iono chromic dye, halo chromic dye, acido chromic dye, chemo chromic dye, electrochromic dye, piezo chromic dye, mechnao chromic dye which changes colour respectively due to UV rays, temperature, polarity of solvents, moisture, ions, pH value, acids, specific chemical agent, electricity, pressure, deformation of substance [11]. Despite of having a large group of these dye only a few one makes a way towards textile. Several factors such as bonding with textiles, end uses and comfort of the wearer etc. are considered while application of smart dyes on textiles. Smart dyes like photochromic dyes and thermochromic dyes, are mainly used in textiles sector. Earlier in 90's, the thermochromism and photochromism were widely used particularly in fashion designing field [12, 13]. Thermochromic dyes have better stability in comparison to photochromic dyes hence these are mostly used dyes in textiles in comparison to other smart dyes [14]. Very few researches related to the use of electrochromic dyes, solvatochromic dyes and mechanochromic dyes on fabric are available. Researches related to the use of electrochromic dyes on textiles are still going on [15]. Different parameter of dye such as maximum absorption, no. of wash cycle, different fastness describes the properties of smart dyes [16]. Application of smart dyes are generally extended to fashion, sport and defence and medical related sector. After application of smart dyes, a simple textile can act as a smart textile which can perform the various function such as Camouflage, thermoregulation etc. [17–19].

Functional dyes can add a specific function to the textile such as protection from UV radiation, microbes as well as insects and moth [20–22]. On the basis of these mentioned function, functional dyes can be categorised into UV protective dyes, antimicrobial dyes and insect and moth repellent dyes. Synthetic and natural dyes can provide these functions but presently more attention is being given to natural dyes due to several environmental concerns. Presently due to climate change issue researches on UV protective dyes and antimicrobial dyes have been increased.

Despite of being constant researches on smart and functional dyes, commercial application is very limited. To widen the application area of smart and functional dyes, it is necessary to use a proper application method to impart these dyes onto textile surface. The knowledge of proper bonding mechanism is important in this regard. The relation between fabric properties and application method should be well known to the researcher. It is also important to work on the light stability and washing stability aspect of these dyes. Researches related to preparation and application of these smart and functional dyes on textiles should be carried out in a significant number. Functional group can be added to a traditionally used dye to enhance its functionality. These kinds of researches would be help in the current scenario and also helpful to increases its commercial application.

2. Types of smart dyes

As discussed in introduction part, only few smart dyes such as photochromic, thermochromic, electrochromic, solvatochromic paved a way for its application in textiles industry. Although researches are still going on to utilise more no. of smart dyes in textile sector. Types of smart dyes that are being used in textiles sector are as follows:

2.1 Electrochromic dyes

Electrochromic dyes are based on electrochromism in which reversible changes of colour occurs due to electricity (gain and loss of electron) [23]. This process occurs generally with some transition metal oxides which conduct both electrically and ions [24]. It is reported that strong electric field can alter the colour of certain dyes [25]. Phthalocyanine dye is a good example of electrochromism [26]. Common electrochromic materials are Polyaniline. In a study, conducting polymer polyaniline layer was formed electrochemically on conducting woven textile substrate which exhibited reversible colour change [27]. During application of electrochromic dyes on textile, performance parameters such as electrochromic contrast, coloration efficiency, write-erase efficiency, switching speed, stability, cycle life, and optical memory are considered. Polyethylene terephthalate (PEPES) membranes were coated with poly-3,4-ethylenedioxythiophene polystyrene sulfonate (PEDOT:PSS) to develop electrochromic textile system [28]. Application of Electrochromic dyes can also be extended to develop smart window curtain which filter sun light accordingly [29]. Other smart application such as display of information or for camouflaging purposes of electrochromic dyes have also been reported [30].

2.2 Thermochromic dyes

Thermochromic dyes causes reversible changes in colour within the absorption spectrum of a thermochromic molecule, usually in the visible light range. These changes are induced by heat. Thermochromic material can be organic, inorganic polymers and sol-gels in nature [31]. Presently, only two type of thermochromic system are used in textiles i.e. liquid crystal type and molecular rearrangement type. Organic leuco dyes also create thermochromic system [32]. Organic thermochromic systems perhaps occur due to equilibrium between molecular species like acid–base, keto-enol, lactim-lactam, stereoisomers or between different crystal structures [33]. Among inorganic thermochromic substances, few examples of metal oxides are known such as indium oxide, zinc oxide, chromium hemitrioxide-alum earth [34]. Application of thermochromic dye on textile can be done using 3 methods which can be used exhaust, continuous method, microencapsulation and printing. In exhaust method thermochromic pigment are applied on to textile with a cationic agent, non-ionic dispersing agent and binder using Material to liquor ratio 1:30 at 70 ° C temperature for half hour [35]. In continuous method thermochromic dye, cationic and non-ionic dispersing levelling agent are applied with acrylic soft binder solution. This solution is applied on textiles through pad dry cure machine. Drying of samples is generally done at 80°C for 3 min while curing is done at 140°C for 2 to 3 minutes [36]. In microencapsulation method, microcapsules were created using colourless dye, precursor and colour developer [37]. These materials melt and solidifies with respective to the application of heat and cold. On melting these material changes its colour, while on solidification come back to its original colour.

Various novel colour effects as well as camouflage designs can be produced using combination of thermochromic, non-thermochromic and mixture of both on the conductive cotton [38]. These thermochromic dyes can take heat from there nearby environment such as sun rays and cause a reversible change in the colour of textiles. These dyes are also establishing their presence in the field of Protective clothing for military [39]. Protective clothing using thermochromic colourant can mimic the colour of its surrounding environment with the change in temperature. Even sometimes colour can change with body temperature. Thermochromic dyes on textiles also used to protect a brand from fake imitation [40]. Microparticles (made of thermochromic and photophoresent dyes) and binder are used to make the brand logo. These logos can be temporary and permanent based on the binder used. The logo made up of thermochromic dye change colour with body temperature [41]. Thermoregulation is also a very imp application of thermochromic dye which is based on heat absorbance (in case of dark colours) and reflectance (in case of light colours). A suitable example of this phenomenon is coating of thermochromic dyes on firemen's uniform on high temperature. The Colour of uniform will be converted to white colour which reflect the heat. Thermochromic dyes can also do thermoregulation via expanding or shrinking the textiles fibre. This Shrinking and expanding of fibres in a fabric causes opening or closing of pores which help regulate the passage of air to the body according to external environment [42].

2.3 Photochromic dyes

Photochromic dyes change its chemical structure due to UV rays. Photochromic effect was first time seen on the tetracene and after that it was observed in potassium salt of dinitomethane [43]. In 1960 first photochromic spiropyrans was developed using printing technique by Ksnebo ltd, Japan. In 1998, photochromic dyes were first time used in textiles to produce camouflage effect. Photochromic dyes also decolorise. This decolourisation is called negative photochromism [44]. Various metal oxides, alkaline earth metal sulphides come under the category of Photochromic dyes. Photochromic dyes are water insoluble hence cannot be applied on wool and cotton [45]. Presently Researches are also going on Water-soluble photochromic dyes. These dyes are relatively cheaper option for textiles and leather. Spirooxazine-based Water-soluble photochromic dyes having sulphonate group have been developed. Sulphonate gp impart water solubility. These water-soluble photochromic dyes are mainly for proteins fibres [46]. The wash fastness and photostability of these dyes are moderate [47].

Application of photochromic dyes can be done on the window curtains and facades. This would be helpful to control sunlight [48]. Heat sensitive photochromic dyes also used as temperature indicator [49, 50]. Dupont company are also working on the camouflaging property of these photochromic dyes to develop camouflaging clothing for armies [51, 52]. Flexible UV sensors based on Photochromic dyes also change its colour due to UV rays which makes a UV detective fabric which help to tell the intensity of UV rays to get protection accordingly [53].

Different application methods have been reported to apply Photochromic dyes and pigments. In a study exhaust method (direct Coloration), photochromic colourants were used with auxiliaries such as dispersing agent and ceramic balls (250 g) and aqueous acetic acid. The dyeing was started at 40° C temperature and temperature raised to 60 ° C and then to 90 ° C [54].

Photochromic colourants as disperse dyes has also been used to dye polyester using exhaust dyeing technique. Photochromic colourants are insoluble in water therefore organic solvents such as Acetone is used to dissolve the photochromic colourant. After dissolution of colourant, dispersing agent is added. Acetic acid is also used to maintain the weak acidity (4.5) of solution. This weak acidity helps to minimise the degradation of colourant. The material to liquor ratio of solution was kept to 1:50. The fabric is dyed at 120 ° C for 45 min. Further, rinsing in cold water was done to reduce the temperature to 70 ° C. This 70 ° C temperature was maintained for 20 minutes with in the solution of sodium hydroxide, sodium dithionite and a non-ionic surfactant. Material to liquor ratio was maintained to 1:30.

Photochromic colourants can also be applied through pad dry cure method. In this method the fabric is dried at 80 ° C for 2 minutes in hot air, followed by 140°C of curing temperature for 3 minutes. Microencapsulation was also used for application of photochromic colourants [55–57].

2.4 Chemo chromic dyes

Chemo chromic dyes change its colour due to differences in pH. Phthalides, fluoranes, triarymethines, and simple azo dyes are the examples of pH sensitive dyes [58].

Conventional exhaust dyeing, solgel method as well as prior addition to electrospinning solution has been reported as methods of application for chemo chromic dyes. Chemo chromic dyes can be used in identify the proteins (of certain microbes) with change of colour. This identification with changing colour indicates the presence of microbes in textiles [59]. Therefore, Application area of Chemo chromic dyes can be widened to medical textiles bandages [60].

2.5 Solvatochromic dyes

Solvatochromic dyes work on the principle of Solvatochromism which corresponds to reversible changes of colour due to solvent polarity. The change in colour is occurred due to shifting in maximum absorption of different solvents [61]. There are two types of solvato chromism i.e. positive solvatochromism and negative solvatochromism. Positive solvatochromism corresponds to a hypsochromic shift induced by a decrease of solvent polarity" and "negative solvatochromism corresponds to a hypsochromic shift induced by an increase of solvent polarity" [62]. Common example Solvatochromic dyes are pyridinium, merocyanine, and stilbazolium dyes. Solvatochormic dyes can be applied in the textiles using microencapsulation technique. Assembly of microsphere on fabric causes colour change on drying and wetting of the fabric. Application of Solvatochromic dyes in textiles is very limited. These dyes can be used to identify the stale or toxic food due to colour change of packaging [63].

S. No	Smart Dyes	External Stimuli responsible for colour change	Suitable methods of application for textiles	Application in textiles	References
1.	Electrochromic dye	Electricity (gain and loss of electron)	Coating	For displaying information or for camouflaging purposes	[23–30]
2.	Thermochromic dye	Heat (absorption spectrum of a thermochromic molecule)	Exhaust, Continuous method, micro encapsulation	Thermoregulation, Brand forgery detection	[31–42]
3.	Photochromic dye	Light	Exhaust and pad dry cure method	Camophlaging textiles, window curtains, UV detective fabric	[43–57]
4.	Chemo chromic dyes	pH	Conventional exhaust dyeing, solgel method, prior addition to electrospinning solution.	Medical textiles	[58–60]
5.	Solvatochromic dye	UV rays	Micro encapsulation	Packaging	[61–63]
6.	Mechanochromic dye	Deformation (elongation and compression) of polymer on which applied	Melt processing technique, physically dispersed in form of supramolecular aggregates in a matrix in a polymer, Covalent insertion of chromophoric units	Footwear and shaped garment industry	[64–66]

Table 1.
Types of smart dyes used in textile industry.

2.6 Mechanochromic dyes

Mechanachromism is a phenomenon in which a polymer changes it colour due to deformation such as elongation and compression. Elongation and compression occur due to the change in pH, temperature. Mechanachromic dyes are the organic dyes which is applied on polymer and causes changes in colour due to certain mechanical pressure. Mechanochormic dye cannot act alone it require a polymer (material) because only polymer can deform (elongate or compress). In a study 1,4-bis (R-cyano-4- methoxystyryl)-2,5-dimethoxybenzene9 (C1-RG, F) has been incorporated in polyethylene using melt processing technique [64]. In another research mechanochromic polymers have been developed by incorporating a dye filled microcapsules [65]. In a study Polydiacetylenes (PDA) was synthesised by thermal polymerisation of diphenyl sulphide containing bisdiacetylene. It is reported in that study that PDA changes its colour from blue to red due to elongation and compression. In the cool (non-extended state) form, the material has a particular colour, this particular colour changes due to mechanical abrasion which causes heating of surface. Mechanochromic colourants can be physically dispersed in form of supramolecular aggregates in a matrix in a polymer. Covalent insertion of chromophoric units into the macromolecule backbone or side chains has also been used to apply mechanochromic colourants [66]. Application area of mechanochromic dyes can be done in footwear and shaped garment industry in which product can change colour due to deformation (**Table 1**).

3. Types of functional dyes

Functional groups like -OH, -NH2, – COOH of dyes offer various functional properties when applied onto the textile [67]. Both natural and synthetic dyes can perform various functions after application on textiles. Now days more emphasise is being given to the green and sustainable functional dyes which directly comes from nature. Types functional dyes are as follows:

3.1 Antimicrobial/antibacterial dye

Microbes and bacteria cause several kinds of dermal infection, body odour and several other severe health issue [68]. Antimicrobial dyes provide the protection from variety of micro-organisms like gram-positive bacteria such as *Staphylococcus aureus, Streptococcus epidermidis and Bacillus cereus* and gram-negative bacteria such as *Escherichia coli, Klebsiella pneumonia, Shigella flexneri and Proteus vulgaris* and other microbes [69]. A large no. of antimicrobial dyes possesses antimicrobial activity against human pathogen but very few have been reported for textiles. Number of researches have been conducting to find out the antimicrobial properties in various synthetic and natural dyes. Various studies reported that synthetic dyes such as direct, cationic, reactive and disperse dyes provides antimicrobial property to the fabric after dyeing. In a study it was found that Direct Blue 168 dye and copper sulphate as mordanting agent incorporate antimicrobial properties in the acrylic fabric [70]. Similarly, antimicrobial dyes based on azo heterocyclic and/or homocyclic Systems also have biocidal behaviour [71]. Multifunctional antimicrobial dyes have also been developed by adding a functional gp (quaternary ammonium salt group) to a traditional aminoanthraquinoid dye [72]. Most research has been focused on polycationic systems that are more supple for modification. Reactive dyes form homopolar bonds with textile substrates. The synthesised thiazolidinone derivatives exhibiting antimicrobial properties. Synthesised monoazo disperse dyes showed better results against gram positive as well as gram negative bacteria [73]. Application of natural dyes on textiles also offer antimicrobial textiles. Phenolic compounds such as anthraquinones, flavonoids, tannins, naphthoquinones and others in natural dyes are responsible for the antimicrobial activity. When these phenolic compounds reacted with textiles, formation of complex form. This complex hinders (bacteriostatic) or kill (bactericidal) the enzyme production in microbes. At present more emphasis is being given to antimicrobial activity of natural dye due to various environmental concern. From various studies it was found that natural dyes extracted from Pomegranate (*Punica granatum)* peels, Henna (*Lawsonia inermis*) leaves, *M. composita* leaves, Madder (*Rubia tinctorium*) root, safflower, *Rumex maritimus (*Golden dock*), Indigo (Quercus infectoria) leaves,* Berberine provides antimicrobial properties to the fabric. Application of Natural dyes such as peony, clove, *Coptis chinensis* (Chinese goldthread) and gall-nut on fabric also provides protection against *Staphylococcus aureus* due to presence of phenolic compounds [74–78].

Perspiration cause formation of bacterial colonies on textiles, which led to bad odour [79]. Various natural dyes can act as a barrier to form these colonies. Natural dyes such as pomegranate, coffe arabica, *Cassia tora*, gardenia Indigo, Peony, clove and pomegranate (*Punica granatum*) reported as a good deodorising agent when applied on textiles [80–82]. Natural dyes extracted from gallnut also act as a deodoriser for textile due to the presence of gallotannin [83].

There is no. of methods to impart antimicrobial dyes on textiles. For proper bonding of textile and dye, the textile surface can be modified through various treatments such as treatment with chemicals, chitosan, enzymes, UV radiation, ultrasound [84]. Application methods can be altered with regard to type of fibre dye and the end use. Therefore, researchers should consider the structure of dye and fibre. The researchers should also have knowledge how this bonding of fibre and dye affect the fabric properties. The products of Health, hygiene as well as medical textiles comes under the application area of antimicrobial dyes.

3.2 UV protective dyes

Presently UV ryas are causing various harmful effects. UV protective dyes enhance the UPF (Ultraviolet Protection Factor) of the textiles. UPF means how much a fabric can protect the wearer from harmful UV rays. In general, all dyes act as a UV absorber because spectral region falls into UV region. Various kind of synthetic dyes are commercially available to enhances the UPF of the fabric. Direct, vat and reactive dyes increases the UPF of fabric [85–87]. Various researches reported that natural dyes can also enhances the UPF of the fabric. Absorption characteristics of natural dyes generally determines the UPF of the fabric [88]. Phenolic compounds in natural dyes work as UV protective agent as these molecules absorb the UV radiation. For instance, *R. maritimus, M. philippinensis, K. lacca, A. catechu and A. nilotica* have tannin content (phenolic compond) thus provide good UPF to the fabric. It is also reported in various studies that, Natural dyes from eucalyptus leaf extract, *Xylocarpus granatum* (*Cedar Mangrove*) bark extract, blossoms of broom (*Cytisus scoparius*) and dandelion (*Taraxacum officinale*), Weld, woad, logwood lipstick tree, madder, brasil wood, and cochineal, gromwell roots, *Acacia*, henna dye extract, chitosan and turmeric dye gallnuts, areca nuts, and pomegranate peels banana peel *babool, ratanjot*, annatto and *manjistha* enhances the UPF of textiles [89–95]. Mordants are used with natural dyes to enhances the fastness properties of the dyes. Several studies reported the positive impact of mordant on the UPF of the fabric. But very few studies also reported the negative impact of mordant on the UPF of the fabric. It means type of the mordant, mordanting method also affect the UPF of the fabric [96, 97]. In case of both synthetic as well as natural dyes, several parameters such as concentration of dye, exhaustion time and extraction and exhaustion temperature affect the UPF of fabric [98, 99]. For instance, it is reported that with the increase of concentration of dye, the UPF of dyed fabric also increases. While exhaustion time and temperature are not causing significant change in the UPF of the fabric. Various studies also report the correlation between the dyeing parameters and UPF.

In a study exhaust method was used for the application of herbal plant extract to enhance the UV protection of the fabric. Madder and cutch dye was applied on nettle fabric using exhaust dyeing method [100]. Pad dry cure method were also used for application of UV protective dye [101].

Application of UV protective dyes can be extended mainly to the clothing of outdoor activities such as fishing, farming, horticulture, gardening, building construction, road construction, postcard distribution, oil production field, military defence services, skiing, police work, professional cycling, surfing [102].

S. No	Functional Dyes	Source	Suitable methods of application for textiles	References
1.	Antimicrobial dyes	**Natural source:** Pomegranate (*Punica granatum)* peels, Henna (*Lawsonia inermis*) leaves, *M. composita* leaves, Madder (*Rubia tinctorium*) root, safflower, *Rumex maritimus (*Golden dock*)*, *Indigo (Quercus infectoria) leaves,* Berberine, peony, clove, *Coptis chinensis* (Chinese goldthread) and gallnut *Synthesised source:* direct, cationic, reactive and disperse dyes, Direct Blue 168 dye and copper sulphate, dyes based on azo heterocyclic and/or homocyclic Systems, developed by adding a functional gp (quaternary ammonium salt group) to a traditional aminoanthraquinoid dye, polycationic systems, Synthesised thiazolidinone derivatives, Synthesised monoazo disperse dyes	Treatment of textiles surface with chemicals, chitosan, enzymes, UV radiation, ultrasound	[70–84]
2.	UV protective dyes	*Natural source: R. maritimus, M. philippinensis, K. lacca, A. catechu and A. nilotica* eucalyptus leaf extract, *Xylocarpus granatum* (*Cedar Mangrove*) bark extract, blossoms of broom (*Cytisus scoparius*) and dandelion (*Taraxacum officinale*), Weld, woad, logwood lipstick tree, madder, brasil wood, and cochineal, gromwell roots, *Acacia*, henna dye extract, chitosan and turmeric dye gallnuts, areca nuts, and pomegranate peels, banana peel, *babool, ratanjot*, annatto and *manjistha* *Synthesised source:* Direct, vat and reactive dyes	Exhaust, pad dry cure and microencapsulation	[85–104]
3.	Moth repellent dyes	Saffron flower waste, onion skin, henna, myrobalan, silver oak leaf, madder, wall nut, dholkanali and yellow roots	Simultaneous dyeing	[105–110]
4.	Mosquito repellent dyes	**Natural source:** Pomegranate peel with polyvinyl alcohol **Synthesised source:** *4-Amino-N, N-diethyl–3-methyl benzamide* (MD).	Exhaust, microencapsulation	

Table 2.
Types of functional dyes used in textile industry.

UV protective dye can also be applied to the clothing of Indoor workers who are potentially exposed to UV radiation for example in hospitals where UV radiation is required for some kind of treatments in some laboratory works, plasma torch operating, printing, lithographing, painting, wood curing, plastic working, in some cases food industry also [103]. Army personnel who have been working in extreme climate conditions also experiences intense solar radiation with terrible heat stress also requires protection from UV rays [104].

3.3 Moth proof and mosquito repellent dyes

Moth proof and mosquito repellent dyes provides protection against moths and mosquitos after application on textiles. Synthetic moth proof or mosquito repellent are generally available in colourless form. Therefore, moth proof and mosquito repellent dyes available in natural form. Various natural dyes contain tannin which can also act as a moth proofing agent. It has been reported in a research that the natural dyes having more than about 40% tannin is effective as an anti-moth agent. In various studies, it was reported that natural dye extracted from Saffron flower waste, onion skin, henna, myrobalan, silver oak leaf, madder, wall nut, dholkanali and yellow roots provides anti-moth properties after application on textiles [105]. Application of dye extracted form pomegranate peel with polyvinyl alcohol can act as mosquito repellent [106]. The mosquito repellent property of synthesised *4-Amino-N, N-diethyl–3-methyl benzamide* (MD) coupled with three different naphthol were assessed. Cotton fabric dyed with MD and naphthol showed very good and durable mosquito repellence. *N, N-diethyl-m-toluamide* (DEET) is used to synthesise the MD [107]. Pomegranate peels Extract can also act as mosquito repellent dye after its application on textiles.

Moth proof dyes were applied using exhaust method in which condition such as (Concentration of colourant-5%, temperature. 90°C degree, M:L- 1;:40, and pH 5–6, were maintained. Simultaneous dyeing with mothproofing agent on wool fabric were also reported. Result of this study showed that undyed and only mordanted fabric provides lesser protection from *D. maculatus* in comparison to madder dyed wool fabric. Mosquito repellent dyes are applied on fabric using either using exhaust method, pad dry cure and microencapsulation methods. [108, 109].

Mothproof and insect repellent dyes is textile museum and library to protect the textiles and books. Mosquito repellent dyes can also be applied to the children's clothing, pram and curtains [110] (**Table 2**).

4. Conclusion

This high technological era is not only based on beautiful products but it is based on "beauty with artificial intelligence". Various researches are being conducted to develop a smart and functional textile which is not only appealing due to its looks but also have an artificial intelligence to give signal according to the change in nearby environment. Dyes can play important role to develop smart and functional textiles. Besides the functionality, sustainable and green aspect of dyeing are also being considered during synthesis and application smart and functional dyes. In comparison to the conventional dyes smart dyes can add a special intelligence to textile fabric such as thermoregulation, camouflaging. Similarly, functional dyes application on textile provides protection form UV rays, unhygienic conditions and insects. Despite of having such an intelligence of to perform according to nearby environment, various smart dyes have lost their ability to develop colour after several molecular transformation. This phenomenon is known as fatigue resistance.

Similarly, functional dyes also have limitation with regards to durability and comfort properties of the fabric. For instance, UV protective dyes have less stability in light and laundry. The fabric dyed with antimicrobial dyes as well as moth and Mosquito repellent also do not have very good wash fastness. These stability-related issue of smart and functional dyes can be enhanced to utilise proper application methods. Different application methods like surface modification (plasma treatment and UV irradiation, etc.) and microencapsulation to enhance the stability of these functional as well as smart dyes. Application method should be such type that could make a balance between comfort properties and stability.

Abbreviations

AI	Artificial Intelligence
UPF	Ultraviolet Protection Factor
UV	Ultraviolet Radiation
PDA	Polydiacetylenes
M:L	Material to liquor ratio

Author details

Deepti Pargai
Department of Clothing and Textiles, G.B. Pant University of Agriculture and Technology, Pantnagar, Uttarakhand, India

*Address all correspondence to: pargai.deepti16@gmail.com

References

[1] Gregory P. Classification of Dyes by Chemical Structure. The Chemistry and Application of Dyes. 1990. pp 17-47

[2] Gürses A, Açıkyıldız M, Güneş K, Gürses S. Classification of Dye and Pigments. Dyes and Pigments 2016. pp 31-45.

[3] Rijavec T. and Bračko S. Smart dyes for medical and other textiles. Materials, Systems and Applications. Woodhead Publishing Series in Textiles. 2007. DOI: 10.1533/9781845692933.1.123

[4] Periyasamy S. and Khanna G. Thermochromic colors in textiles Americos Industries India. 2008. Retrieved from www.americosind.com

[5] Ferrara M. and Bengisu M. Intelligent design with chromogenic materials. Journal of the International Colour Association (2014): 3, 54-66

[6] Somani P R. Chromic Materials, Phenomena and their Technological Applications. Applied Science Innovations Pvt. Ltd. 650p.

[7] Gong X, Hou C, Zhang Q, Li Y and Wang H. Solvatochromic structural color fabrics with favourable wearability properties. Journal of Material Chemistry.8. 2008.

[8] Ibrahim W. An Investigation into Textile Applications of Thermochromic Pigments. PhD thesis

[9] Morsümbül S and Akumbasar P. Photochromic textile materials. IOP Conference Series Materials Science and Engineering. 2018. 459(1):012053 DOI: 10.1088/1757-899X/459/1/012053.

[10] Kelly F M and Cochrane C Handbook of Smart Textiles Color-Changing Textiles and Electrochromism. pp 859-889

[11] Christie R M. Chromic materials for technical textile applications. In book: Advances in the Dyeing and Finishing of Technical Textiles 2013 DOI: 10.1533/9780857097613.1.3

[12] Morsümbül S and Akçakoca E P and Kumbasar IOP Conference Series: Materials Science and Engineering, Volume 459, Aegean International Textile and Advanced Engineering Conference (AITAE 2018)5-7 September 2018, Lesvos, Greece

[13] Coghlan A. Technology: Clothes that change colour in the heat of the moment. New scientist. 1991. Retrieved from https://www.newscientist.com/article/mg13017684-900-technology-clothes-that-change-colour-in-the-heat-of-the-moment/

[14] Chowdhury M A, Joshi M and Butola B S Photochromic and Thermochromic Colorants in Textile Applications. March 2014Journal of Engineered Fibers and Fabrics 9(1):107-123. DOI: 10.1177/155892501400900113

[15] Sheng M, Zhang L, West J L and Fu *S. multicolor* Electrochromic Dye-Doped Liquid Crystal Yolk–Shell Microcapsules *ACS Appl. Mater. Interfaces* 2020, 12, 26, 29728-29736.

[16] Abate M T. Seipela S, Yu J, Viková M, Vik M, Ferri A, Guanc J, Chen G, Nierstrasza V. Supercritical CO2 dyeing of polyester fabric with photochromic dyes to fabricate UV sensing smart textiles. Dyes and Pigments. 2020. Volume 183

[17] Karpagam K R, Saranya K S, Gopinathan J and Bhattacharyya A. Development of smart clothing for military applications using thermochromic colorants. Journal of the Textile Institute. 108: 7. 2016. 755-765p

[18] Baumbach J. Colour and camouflage: design issues in military clothing,

in Advances in Military Textiles and Personal Equipment. Woodhead Publishing Series in Textiles 2012. Pages 79-102

[19] Urquhart J. Smart textile uses sweat as switch to keep wearer cool or warm royal society of chemistry. Royal society of chemistry. 2019. Retrieved from https://www.chemistryworld.com/news/smart-textile-uses-sweat-as-switch-to-keep-wearer-cool-or-warm/3010099.article

[20] Reda M. and El-Shishtawy. Functional Dyes, and Some Hi-Tech Applications. 2009. https://doi.org/10.1155/2009/434897.

[21] Mishra V R, Ghanavatkar C W, Sekar N. UV protective heterocyclic disperse azo dyes: Spectral properties, dyeing, potent antibacterial activity on dyed fabric and comparative computational study. Spectrochimical Acta Part A: Molecular and Biomolecular Spectroscopy.2019. 223.117353

[22] Mohamed F A and Ibrahim H M. Antimicrobial dyes based on heterocyclic and/or homocyclic systems for dyeing and textile finishing Textile Research Division, National Research Center. 2014. 8(8), .285-301.

[23] Gregory P. Electrochromic Dyes. In: High-Technology Applications of Organic Colorants.1991. Topics in Applied Chemistry. Springer, Boston, MA. https://doi.org/10.1007/978-1-4615-3822-6_7 pp 53-56

[24] Granqvist, C. G Electrochromic Metal Oxides: An Introduction to Materials and Devices https://application.wiley-vch.de/books/sample/3527336109_c01.pdf

[25] John R. Platt. Electrochromism, a Possible Change of Color Producible in Dyes by an Electric Field. 1961. J. Chem. Phys. 34, 862 https://doi.org/10.1063/1.1731686

[26] Mortimer R J. Switching Colors with Electricity. American Scientist. 2013. 101(1):38. DOI: 10.1511/2013.100.38

[27] Granqvist C J Energy-Efficient Windows: Present and Forthcoming Technology in Materials Science for Solar Energy Conversion Systems. 1991

[28] Graßmann C, Mann M, Langenhove L V and Pfeiffer, A. S. Textile Based Electrochromic Cells Prepared with PEDOT: PSS and Gelled Electrolyte

[29] Michaelis A, Berneth H, Haarer D, Kostromine S, Neigl R, Schmidt R. Electrochromic Dye System for Smart Window Applications. Advanced material.2001. 13(23). 1825-1828.

[30] Yablonovitch E. Electrochromic Adaptive Infrared Camouflage. Interim Progress Report U.S. Army Research Office DAAD University of California, Los Angeles. Period covered: August 1999 – January 2005.

[31] S. Periyasamy and Khanna G. Thermochromic colors in textiles. 2008

[32] Strižić M, Suzana J, Preprotic P, Gunde M K. Dynamic Colour Changes of Thermochromic Leuco Dye and Liquid Crystal Based Printing Inks. Conference: 4th International Joint Conference on Environmental and Light Industry Technologies. 2013. DOI: 10.13140/RG.2.2.34274.56005

[33] Ajeeb F, Younes B, Khsara A K. Investigating the Relationship between Thermochromic Pigment Based knitted Fabrics Properties and Human Body Temperature

[34] Lieva Van Langenhove(ed.) Smart Textiles for Medicine and Healthcare: Materials, Systems and Applications

[35] Matsuda K and Irie M. Chemistry Letters. 2006. 35: 1204

[36] Chowdhury, M.A.; Butola, B. and Joshi. M. Application of thermochromic

colorants on textiles: Temperature dependence of colorimetric properties. Coloration Technology. 2013. 129(3) DOI: 10.1111/cote.12015

[37] Aksoy S A and Alkan C. January 2, 2018. Microencapsulation of Three-Component Thermochromic System for Reversible Color Change and Thermal Energy Storage,

[38] Chowdhury M A, Butola B and Joshi M. Development of Responsive Camouflage Textile using Thermochromic and Non-thermochromic Colorants. 2013

[39] Langenhove L V. Smart Textiles for Medicine and Healthcare: Materials, Systems and Applications.

[40] Americos thermochromic microcapsule. Retrieved from https://www.kenencoregroup.com/smart-colorants.html

[41] Tebbe, G. Textile material for garments. United States Patent Application Publication US 2002/0137417 A1. 2002-09-26.

[42] Hibbert, R. Textile innovation, 2002. London.

[43] Durr, and Bouas-Laurent, H, Pure Applied Chemistry, 2001; 73; 639.

[44] Aiken S, Edgar R, Gabbutt C, Heron B M, Hobson P A. Negatively photochromic organic compounds: Exploring the dark side. Dyes and Pigments.2018.149:92-121.

[45] Billah R, Christie R M, Shamey, R Direct coloration of textiles with photochromic dyes. Part3: Dyeing of wool with photochromic acid dyes. 2012. Review of Progress in Coloration and Related Topics 128(6):488-492. DOI: 10.1111/j.1478-4408.2012.00406.x.

[46] Billah R, Christie R M, Shamey R M. Direct coloration of textiles with photochromic dyes. Part 1: Application of spiroindolinonaphthoxazines as disperse dyes to polyester, nylon and acrylic fabrics. 2008. Coloration Technology 124(4):223-228. DOI: 10.1111/j.1478-4408.2008.00145.x

[47] Shah M R B Photochromic protein substrates, Mol Cyst Liq Cryst, 2005. 431, 235/[535]-239/[539].

[48] Addington D M and Schodek D L . Smart materials and new technologies, Amsterdam, Elsevier.2005.

[49] F Fu and L Hu. Temperature sensitive colour-changed composites, in Advanced High Strength Natural Fibre Composites in Construction, 2017.

[50] Zhang Y, Hu Z, Xiang H, Zhai G, Zhu M. Fabrication of visual textile temperature indicators based on reversible thermochromic fibers. Dyes and Pigments. 2019. 162: 705-711

[51] Santos L D M, Townes D E, Patricio G R, Winterhalter C A, Dugas A, O'Neill T R, Lomba R A, Quinn B J. Camouflage U.S. Marine corps utility uniform: pattern, fabric, and design. US patent. 2001

[52] ALDIB M. An Investigation of the Performance of Photochromic Dyes and their Application to Polyester and Cotton Fabrics pHD Thesis. Watt University. Scottish Borders Campus School of Textiles and Design. 2013

[53] Viková M, and Vik M Description of photochromic textile properties in selected color spaces 2014

[54] Aldib M and Christie R M. School of Textiles & Design. Textile applications of photochromic dyes. Part 4: application of commercial photochromic dyes as disperse dyes to polyester by exhaust dyeing

[55] Little A F and Christie, R *M. textile* applications of photochromic dyes. Part

1: Establishment of a methodology for evaluation of photochromic textiles using traditional colour measurement instrumentation April 2010 Coloration Technology 126(3):157-163 DOI: 10.1111/j.1478-4408.2010.00241.x

[56] Aldib M and Christie R *M. textile* applications of photochromic dyes. Part 5: application of commercial photochromic dyes to polyester fabric by a solvent-based dyeing method. Coloration technology. 2013. Volume129, Issue2. Pages 131-143

[57] Topbas O, Sariisik A M, Erkan, G and Ek O Photochromic microcapsules by coacervation and in situ polymerization methods for product-marking applications. Iranian Polymer Journal. 2020. 29: 117-132.

[58] Christie R M. Chromic materials for technical textile applications. In Advances in the Dyeing and Finishing of Technical Textiles; Woodhead Publishing: Oxford, UK; Cambridge, UK; Philadelphia, PA, USA; New Delhi, India, 2013; pp. 3-36.

[59] Stojkoski V and Kert M. Design of pH Responsive Textile as a Sensor Material for Acid Rain. Polymers (Basel). 12(10): 2251. 2020 doi: 10.3390/polym12102251

[60] 内森·扎马利帕杰夫·葛雷. Chemochromic medical articles. 2014 - Boston Scientific Scimed, Inc

[61] Nigam S and Rutan S. Principles and Applications of Solvatochromism Applied Spectroscopy 2001. 55(11):362-DOI: 10.1366/0003702011953702

[62] Papadakis R and Tsolomitis A. Synthesis, substituent and solvent effects on the UV-Vis spectra of 4-pentacyanoferrate-4′-aryl substituted bipyridinium complex salts. 2008. Conference: ICPOC 2008 - 19th IUPAC Conference on Physical Organic ChemistryAt: Santiago de Compostela Spain

[63] Gong X, Hou C, Zhang Q, Li Y and Wang H. Solvatochromic structural color fabrics with favorable wearability properties. Journal of Material chemistry.

[64] Crenshaw B R, Burnworth M, Khariwala D, Hiltner A, Mather P T, Simha R, Weder C. Deformation-Induced Color Changes in Mechanochromic Polyethylene Blends. Macromolecules. 2007. 40(7) DOI: 10.1021/ma062936j

[65] Calvino C, Henriet E Li, Muff L F, Schrettl S, Weder C. Mechanochromic Polymers Based on Microencapsulated Solvatochromic Dyes. 2020Macromolecular Rapid Communications 41(7) DOI: 10.1002/marc.201900654

[66] Ciardelli F, Giacomo R, Pucci A. Dye-Containing Polymers: Methods for Preparation of Mechanochromic Materials. 2012. Chemical Society Reviews 42(3) DOI: 10.1039/c2cs35414d

[67] Vankar, P. Chemistry of Natural Dyes. RESONANCE. 2000. 73-80 pp. https://www.ias.ac.in/article/fulltext/reso/005/10/0073-0080

[68] Aly R, Baron S (ed.). Microbial Infections of Skin and Nails. In: Medical Microbiology. 4th ed. Galveston (TX): University of Texas Medical Branch at Galveston; 1996. Chapter 98.

[69] Lowy F. Bacterial classification, structure and function. 2009;44(12): 977-983.

[70] Yusuf M, Shabbir M, Mohammad F. Natural colorants: Historical, processing and sustainable prospects. Natural Products and Bioprospecting. 2017;7(1): 123-145. DOI: 10.1007/s13659-017- 0119-9.

[71] F. A. Mohamed, H.M. Ibrahim. researches and reviews in bio sciences volume 8 issue 8. Antimicrobial

dyes based on heterocyclic and/or homocyclic systems for dyeing and textile finishing

[72] Sun G and Ma M. Multifunctional antimicrobial dyes. Patent no. US20050011012A1. University of California University of California San Francisco UCSF.

[73] Abedi D, Mortazavi S M, Mehrizi M K, Feiz M. Antimicrobial Properties of Acrylic Fabrics Dyed with Direct Dye and a Copper Salt. 2008. Textile Research Journal 78(4):311-319. DOI: 10.1177/0040517508090486.

[74] Mohammed S, Morsy A M, Apasery A E. Huda Mahmoud Disperse Dyes Based on Aminothiophenes: Their Dyeing Applications on Polyester Fabrics and Their Antimicrobial Activity Molecules. 2013 18(6):7081-92 DOI: 10.3390/molecules18067081.

[75] Rehman F, Sanbhal N, Naveed T, Farooq A, Wang Y, Wei W. Antibacterial performance of Tencel fabric dyed with pomegranate peel extracted via ultrasonic method. Cellulose. 2018. 25(7). DOI:10.1007/s10570-018-1864-6

[76] Yusufa M, Shahida M, Khan M I, Khan S A, Khan M A. Mohammed, F. Dyeing studies with henna and madder: A research on effect of tin (II) chloride mordant. *Appl Environ Microbiol.* 2014. 80(21): 6611-6619.

[77] Pal A, Tripathi Y C, Kumar R, Upadhyay L. Antibacterial Efficacy of Natural Dye from Melia composita Leaves and Its Application in Sanitized and Protective Textiles Journal of Pharmacy Research 2016. 10(4):154-159

[78] Lee Y H, Hwang E K, Baek Y M, Kim, H D. Colorimetric Assay and Antibacterial Activity of Cotton, Silk, and Wool Fabrics Dyed with Peony, Pomegranate, Clove, Coptis chinenis and Gallnut Extract Materials. Fibers and Polymers. 2009. 17 (4):560-568

[79] Teufel L, Pipal A, Schuster K C, Staudinger T, Red B. Material-dependent growth of human skin bacteria on textiles investigated using challenge tests and DNA genotyping. Journal of Applied Microbiology 2009108(2):450-61 Follow journal DOI: 10.1111/j.1365-2672.2009.04434.x.

[80] Pargai D, Jahan S and Gahlot M. Functional Properties of Natural Dyed Textiles. In book: Chemistry and Technology of Natural and Synthetic Dyes and Pigments. Intechopen. 2020.

[81] Hwang E K, Lee Y H, Kim H D. Dyeing, fastness, and deodorizing properties of cotton, silk, and wool fabrics dyed with gardenia, coffee sludge, *Cassia tora*. L., and pomegranate extracts. Fibers and Polymers 2008. 9(3):334-340. DOI: 10.1007/s12221-008-0054-9

[82] Koh E and Hong K H. Gallnut extract-treated wool and cotton for developing green functional textiles. Dyes and Pigments 2014103:222-227 DOI: 10.1016/j.dyepig.2013.09.015

[83] Lee Y H, Hwang E K, Baek Y M, Kim, H D. Colorimetric Assay and Antibacterial Activity of Cotton, Silk, and Wool Fabrics Dyed with Peony, Pomegranate, Clove, Coptis chinenis and Gallnut Extract Materials. Fibers and Polymers. 2009. 17 (4):560-568

[84] Erkan G, Şengül K and Kaya S. Dyeing of white and indigo dyed cotton fabrics with Mimosa tenuiflora extract April 2014 Journal of Saudi Chemical Society 18(2). DOI: 10.1016/j.jscs.2011.06.001

[85] Kan C W and Au C H. Effect of direct dyes on the UV protection property of 100% cotton knitted

fabric. Fibers and Polymers volume 16, pages1262-1268(2015)

[86] Bajaj P, Kothari V K, Ghosh S B. Some Innovations in UV Protective Clothing", Indian J. of Fibres and Textile Research 2000 35 (4) 315-329

[87] Wong W Y, Lam J K C and Kan C W. Influence of reactive dyes on ultraviolet protection of cotton knitted fabrics with different fabric constructions. 2015. https://doi.org/10.1177/0040517515591776

[88] Gupta D, Ruchi. UPF characteristics of natural dyes and textiles dyed with them. 2007. Colourage 54(4):75-80.

[89] Pisitsak P, Hutakamol J, Jeenapak S, Wanmanee P, Nuammaiphum J and Thongcharoen R Natural dyeing of cotton with Xylocarpus granatum bark extract: Dyeing, fastness, and ultraviolet protection properties. Fibers and Polymers. 2016. 17 (4):560-568.

[90] Křížová and Wiener. Comparison of UV Protective Properties of Woollen Fabrics Dyed with Yellow Natural Dyes from Different Plant Sources. Environmental Sciences. 2016. 2 (7):2454-9916

[91] Griffoni D, Bacci L, Zipoli G, Carreras G, Baronti S. and Sabatini F. Laboratory and outdoor assessment of UV protection offered by flax and hemp fabrics dyed with natural dyes, Photochem. Photobiol. 2009.85: 313-320

[92] Hong K, Bae J H, Jin S R, Yang J S. Preparation and properties of multi-functionalized cotton fabrics treated by extracts of gromwell and gallnut Cellulose 2011. 19(2): 507-515. DOI: 10.1007/s10570-011-9613-0

[93] Alebeida O K, Taoa Z, Seedahmedc A. New Approach for Dyeing and UV Protection Properties of Cotton Fabric Using Natural Dye Extracted from Henna Leaves. Fibres & Textiles in Eastern Europe. 2015. 23(5):61-65

[94] Orabodee S, Chotima S, Jantip S, Potjanart S, Porntip S B. Effect of Chitosan and Turmeric Dye on Ultraviolet Protection Properties of Polyester Fabric. Applied Mechanics & Materials Academic Journal. 2014. 535: 658.

[95] Jung, J.S. Study of Fastness, UV Protection, Deodorization and Antimicrobial Properties of Silk Fabrics Dyed with the Liquids Extracted from the Gallnuts, Areca Nuts, and Pomegranate Peels. MATEC Web of Conferences. 2016.49:1-6

[96] Gawish SM, Mashaly HM, Helmy HM, Ramadan AM and Farouk R. 2017. Effect of Mordant on UV Protection and Antimicrobial Activity of Cotton, Wool, Silk and Nylon Fabrics Dyed with Some Natural Dyes. Journal of Nanomedicine & Nanotechnology. 8:1

[97] Pargai D, Gahlot M, Rani A. Ultraviolet protection properties of nettle fabric dyes with natural dyes. Ind. J. Fibre Text Res. 2016. 41(4): 418-425

[98] Pargai D and Jahan S. Direct Application of *Vitis vinifera* (Grape) Leaves Extract on Cotton Fabric: A Potential to Prevent UV Induced Skin Problems. Current World Environment 2018. 13(1):165-171 Follow journal DOI: 10.12944/CWE.13.1.16

[99] Osterwalder U, Rohwer H. Improving UV Protection by Clothing — Recent Developments 2002. cancer research160:62-9 DOI: 10.1007/978-3-642-59410-6_9

[100] Sarkar. A K. An evaluation of UV protection imparted by cotton fabrics dyed with natural colorants

BMC Dermatol. 2004; 4: 15. doi: 10.1186/1471-5945-4-15

[101] Mongkholrattanasit R. UV protection properties of silk fabric dyed with eucalyptus leaf extract. The Journal of The Textile Institute Volume. 2011.102 (3)

[102] Adams J. Research on Pad-Dry Dyeing and Ultraviolet Protection of Silk Fabric Using Dyes Extracted from Laccifer lacca Kerr. Advanced Materials Research 2014. 1010-1012:512-515. DOI: 10.4028/www.scientific.net/AMR.1010-1012.512 'Sun-protective Clothing', *Journal of Cutaneous Medicine and Surgery*, 1998.3(1): 50-53

[103] Schmidt. Textile Sun Protection for Soldiers in Extreme Climates. 2010. Retrieved from http://www.army-technology.com/contractors/personal/hohenstein/press1.html

[104] Zajtchuk R, Military Dermatology. Office of The Surgeon General. Department of the Army, United States of America. Retrieved from https://fas.org/irp/doddir/milmed/milderm.pdf

[105] Shakyawa D B, Raja ASM, Kumar A, Pareek P K. Antimoth finishing treatment for woollens using tannin containing natural dyes. Indian journal of fibre and textile research 2015. 40. 200-202p

[106] Rimpi and Singh A. Protection and application of natural mosquito repellents cotton fabric through dyeing

[107] Teli M D. Dyeing of cotton fabric for improved mosquito repellency. 2017. 427-434.

[108] Sajib M I, Banna B U, Mia R, Ahmed B, Chaki R, Alam S S, Rasel M A, Tanjirul Mosquito repellent finishes on textile fabrics (woven & knit) by using different medicinal natural plants 6:4 : 2020

[109] Thite A G, Gudiyawar M Y. Development of Microencapsulated Ecofriendly Mosquito Repellent Cotton Finished Fabric By Natural Repellent Oils. International journal of sciences technology and management.

[110] Nazari A and Branch Y. Efficient mothproofing of wool through natural dyeing with walnut hull and henna against *Dermestes maculatus* 2016. 755-765

Chapter 7

Applications of Metal Complexes Dyes in Analytical Chemistry

Mariame Coulibaly

Abstract

Trace elements, especially heavy metals, are considered to be one of the main sources of pollution in the environment since they have a significant effect on ecological quality. Commonly, the analytical methods for the determination of trace metals are the spectrometry techniques. While, the electroanalytical methods are recognized as a powerful technique for trace metals owing to its remarkable sensitivity, relatively inexpensive instrumentation, ability for multi-element determination at trace and ultra trace level. New alternative electrode materials are highly desired to develop sensitive stripping sensors for meeting the growing demands for on-site environmental monitoring. Dyes aromatic heterocyclic compound, used in food, textile and cosmetic industries has been used for spectrophotometric determination of metals. In electrochemitry, methods for metals determination based on their complexation with dyes were proposed. In this chapter, a brief summary of spectrometry methods and electrochemical sensors for heavy metals detection based on the formation of metals dyes complexes is presented.

Keywords: heavy metals, dyes, dye/complexation, electrochemical analysis, metal complex dye, spectrometry

1. Introduction

Dyes are known to be used in the textile industry, printing, food industry, as well as cosmetic industry. Since the invention of synthetic dyes in 1856, chemistry has been enriched by these large group of chemical compounds. More than 800,000 tons are manufactured by year [1–4]. Dyes are organic substances with chromophore and auxochromic groups (**Figure 1**) classified into several groups (indigoid dyes, xanthene dyes, etc.) of various structures for different applications.

These applications depend on dyes chemical structure, their hue and their entire light absorbing system. The chromophores are the groups of atoms responsible for the dye colour and auxochromes are an electron withdrawing or donating substituents that cause or intensify the colour of the chromophores [6] in shifting the adsorption towards longer wavelength along with an increase in the intensity of absorption. Some commonly known chromophores groups are: azo (–N=N–), carbonyl (–C=O), methine (–CH=, nitro (–NO2) and quinoid groups. The auxochromes are acids or bases; the most important are amine (–NH3), carboxyl (–COOH), sulfonate (–SO3H) and hydroxyl (–OH).

The use of dyes in analytical chemistry is well known. Dyes applications in analytical chemistry are feasible because of the presence of chromophores and auxochromes [7]. Most of dyes form complexes with pollutans in aqueous media [8].

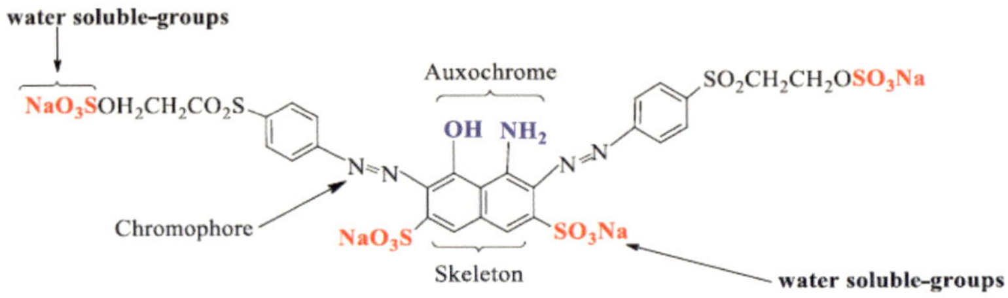

Figure 1.
Structure of the azo reactive dye [5].

They are used as titrations indicator in analytical chemistry, and their complexes with the metal ions in aqueous media are used in spectrophotometric analysis. The complexation between dyes and some essential metals including Cu(II), Hg(II) allows the detection of them by the spectrophotometric method or cromatography [9–11].

As in spectrophotometry, titration, colorimetry or chromatogarphy, dyes are also used in electrochemistry specially in metal ions detection. Electrochemical analysis is recognized to be a method for industrial process control, environmental monitoring, and different applications in medicine [12–16]. The electrochemical technique, especially stripping voltammetry for the trace analysis of metal ions, obtained considerable interest because of its low cost, easy operation, good sensitivity, high selectivity and accuracy [17]. The usual working electrode for stripping voltammetry was a mercury electrode [18] and bare electrodes. However mercury is toxic and causes harm to the environment and human bodies. Concerning bare electrodes, they have numerous limitations such as poisoning, low sensibility, poor stability. Therefore, many groups tried to develop mercury free-electrodes and modified electrode to determine metal ions by voltammetric analysis [19–22]. These chemically modified electrodes (CME) have received an increasing attention in recent years in the fields of electroanalysis due to well recognized advantages in comparison with conventional electrodes [23]. Several reagents and techniques are used to modify the electrodes surfaces [24–28]. The complexation reactions with organic or inorganic reagents on electrodes surfaces, incorporate in electrode paste or in solution for electrochemical anaysis have been reported [29–32]. Among them, the dyes which are important complexing agents for metals. Electrochemical methods for metal ions determination based on their reaction with dyes, their complexation with dye film on electrode surface or inside of electrode paste have been studied [33–35].

2. Metal-complex dyes in spectrophotometric analysis

2.1 Overview

Many studies have been based on the spectrophotometric determination of metal ions after their reactions with complexing reagents including dyes [36–38]. The dyes are organic substances with chromophore and auxochrome groups which can be classified in different type. Among these different type of dyes, azo dyes represent the largest production volume. They make up about 70% of all synthesized dyes annually [39]. The importance of these may increase in the future and also their use in a variety of applications such as complexing agent in spectrophotometric analysis. However, their stability causes environmental pollution once the dyes are discharged with liquid effluents without adequate treatment before release into the natural environment.

The formation of dye complex depends of the number of ligands in the dye structure, and the coordination number of the metal. The electron donating ligand or ion combines with the metal ion to form the complex. For instance, for copper ion which is a bivalent ion with coordination number four, it can complexed with two bidentate ligands in an acid dye or a trivalent or a tetravalent one [40]. Metal-complex dyes formed may be broadly divided into two classes: 1:1 metal complexes and 1:2 metal complexes [41]. These complexes have versatile application in various fields include the dyeing of nylon and protein fibers, paint, toners for photocopiers, laser and ink-jet printers, photoconductors for laser printers, nonlinear optics, singlet oxygen generators, dark oxidation catalysts, and high-density memory storage devices [42]. Their colors span the entire spectrum allowing their use in spectrometry.

By UV–Vis Spectrophotometry, the absorption spectra of solutions allows the determination of metals concentration. The absorption spectra of dye soltution and metal ions solutions are measured first Then, after the mix of the dye and the metal ion, the formation of coloured complex between the both compound give a new color peaking and a new absorption spectra.

2.2 Spectrophotometric determination of trace metal by formation of complexes with dye

It well know that dyes can form a stable complexes with metal ions. Dyes applications in spectrophotometric analysis are possible because of the presence of chromophores and auxochromes (**Figure 2**) [7]. Dyes especially the azo dyes are used as spectrophotometric chemosensor. These compounds interact easily with metal ions through the heteroatoms S, N, and O and can chelate with a large number of metal ions to form a metal-dye complex (**Figure 3**). Numerous works have been dedicated to the synthesis and spectral characterization of new azo dyes and their metal complexes [44, 45]. These studies allow to establish the optimal conditions of formation of the complexes (ratio metal: dye, pH, temperature, the maximum light absorption, the influence of foreign ions ...) and the determination of the constants of complexes.

The complex formation equilibrium and formation constant of the complex can be represented by Eqs. (1) and (2) [46].

$$M + nL \rightarrow ML_n \tag{1}$$

$$K = \frac{[ML_n]}{[M][L]^n} \tag{2}$$

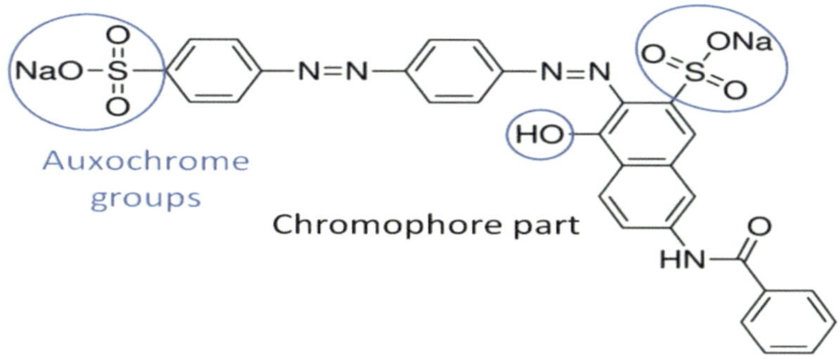

Figure 2.
The light absorption system of dyes [43].

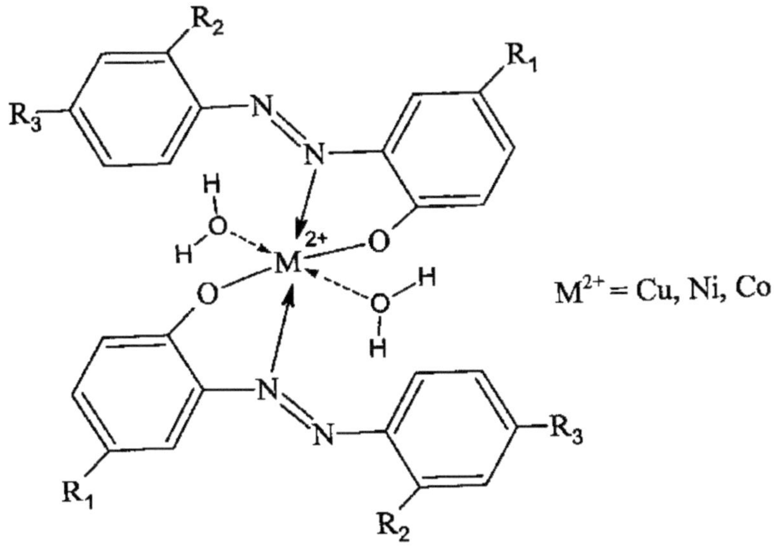

Figure 3.
Chemical structure of the azo-metal chelates [44].

[M], [L] and [MLn] represent the molar equilibrium concentrations of the metal ion, ligand dye and the complex, respectively.

Thus, several new spectrophotometric methods for determination of metal ions based on their complexation with dyes have been developed and tested in real samples [47–49]. Bonishko et al. [48] have developed, a simple spectrophotometric method for the determination of osmium (IV) ions, based on the formation of a complex of this metal with Congo Red. While, the orange G has been used as a complexing reagent in spectrophotometric determination of osmium(IV) by Rydchuk et al. [49]. They showed that the optimum conditions for the formation of coloured complex compound between Os(IV) and acidic monoazo dye Orange G (OG) were: the stoichiometric ration in the complex was 2:1 at pH = 5.80. Moreover, their study showed that at the room temperature Os(IV) practically did not interact with OG. Os(IV)-OG compound was almost fully obtained after 30 min of heating on a boiling water bath (~98°C).

In general, the formation of complexes lead a significant decreases in the absorption band of the dyes and the emergence concomitantly of a new absorption band with different absorbance. Thus, the formation of complex species between mercury and indigo carmine((Hg)IC and (Hg)2IC) allowed a optical determination of mercury [36]. The interaction between Cu(II) ions and indigo carmine forms Cu2(IC) complex characterized by the stoichiometric ratio between indigo carmine and copper 2:1, the molar absorptivity 1.17×10^4 mol L^{-1} cm^{-1} at 715 nm and the stability constant of the complex log K = 5.75, at pH 10, obtained by spectrophotometric data. This complex has been successfully tested for determination of copper in pharmaceutical compounds [37].

3. Electrochemical method for the determination of trace metal by formation of complexes with dye

3.1 Electrochemical behaviour of dye

The electrochemical behaviour of dyes depends of their characteristics, the working electrodes and the pH of supporting electrolyte. According nature of electrodes, the voltammograms of dyes exhibited irreversible oxidation peaks [50] or can be involved in a two or more steps redox reaction [51]. The voltammetric response of indigo carmine shows two well separated peak pairs on graphite electrode (**Figure 4**) at pH 7, while the first pair of peak disapear at ph more basic.

As indicated previously, azo dyes are an important class of organic dyes which consist of at least a conjugated chromophore azo (–N=N–) group. This is the largest and most versatile class of dyes. These dyes are characterised by the presence in their molecules of one or more azo groups —N=N— which form links with organic groups, of which at least one is usually an aromatic nucleus (**Figure 2**). Taking account their potential toxicity, electrochemical methods was developed for the analyzing of azo dye. The mechanism based on the reduction of the azo group with a classical dropping mercury electrode or static mercury drop electrode has been described in detail [52]. Recently, several modified electrodes have been used to study the electrochemical characteristics of azo dyes and their electrochemical determination [54–57]. On a glassy carbon modified, the voltammograms exhibited a irreversible oxidation peaks and a well-resolved oxidation wave was observed at approximately 0.74 V for the azo dye sudan I, sudan II, sudan III, and sudan IV [50] and similar irreversible oxidation peaks was obtained with the congo red on graphene oxide modified electrode [56]. However, the release potential of anodic peak depends of dye and electrode.

These studies show that some of azo dyes are electrochemically reactive. They can reduced or oxidized on different bare or modified electrode (**Table 1**). These electrochemical behaviour allows the detection of dye by voltammetric technique but also the detection of trace metals based on the decrease of dyes oxidation/reduction peak after their complexation.

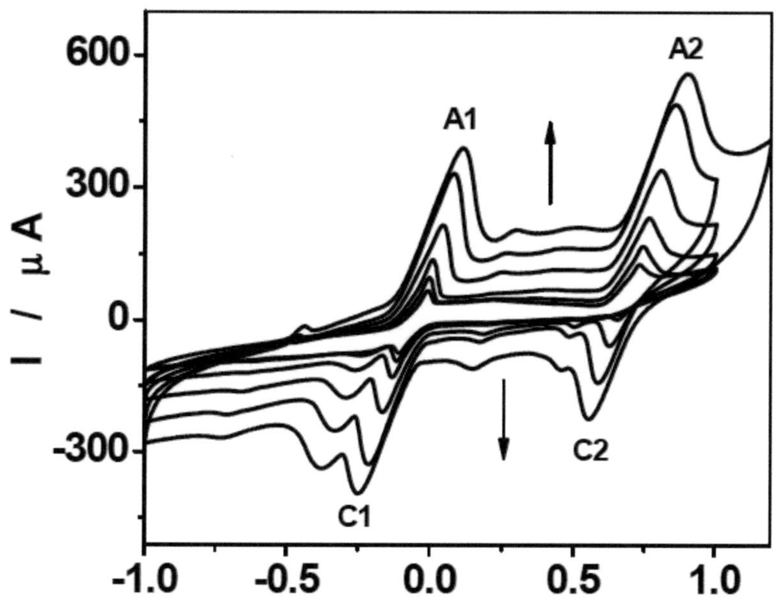

Figure 4.
Cyclic voltammograms recorded at graphite electrode for 10–3 M IC. Experimental conditions: supporting electrolyte 0.1 M phosphate buffer (pH 7); start potential, –1.0 V vs. Ag/AgCl,KClsat; scan rates: 25, 50, 100, 250, 500 and 750 mV/s. [51].

Electrode	Dye/structure	Electrochemical behaviour	Reference
Graphite	Indigo carmine 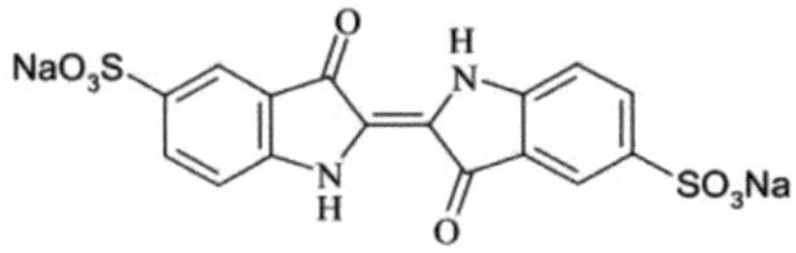	Two pairs reversible oxidation and reduction	[51]
Nanoclay modified carbon electrode	Sunset yellow 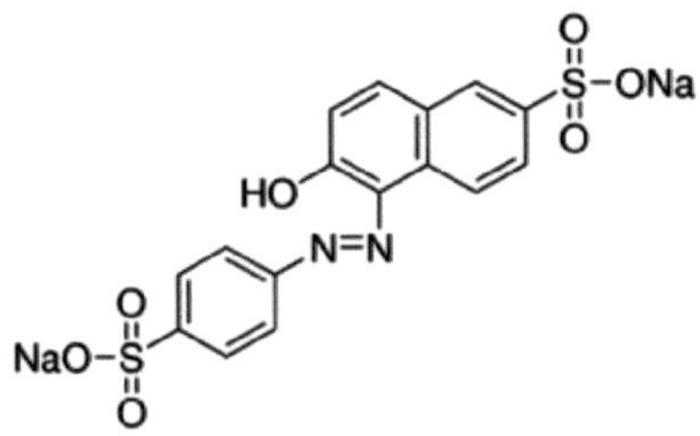	Irreversible oxidation	[58]
Amalgam electrodes	Amaranth 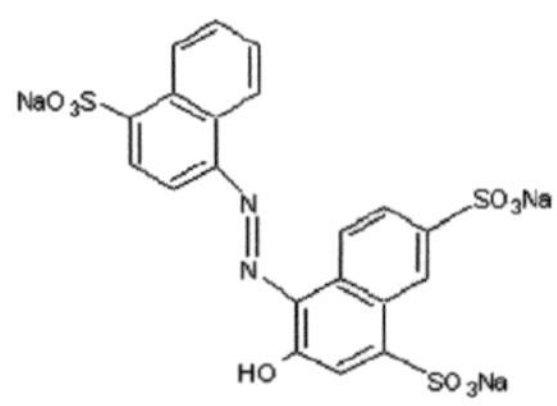	Irreversible reduction	[52]
Amalgam electrodes	Allura Red AC 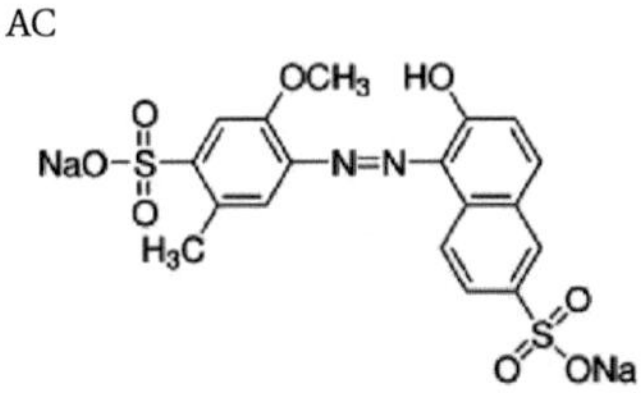	Irreversible reduction	[52]
Graphene oxide casted glassy carbon electrode	Congo red 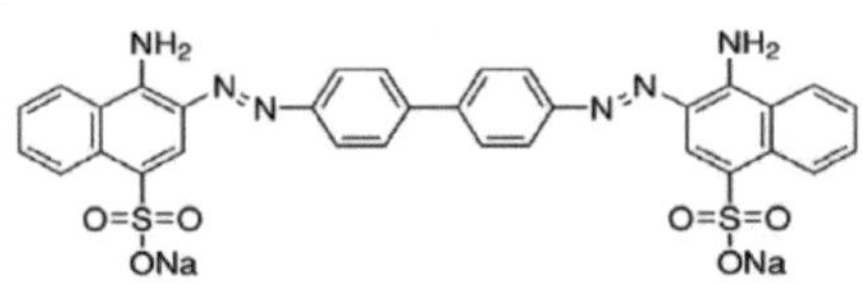	Irreversible oxidation	[56]
Carbon nanotube ionic liquid gel modified glassy carbon	Sudan I, II, III and IV 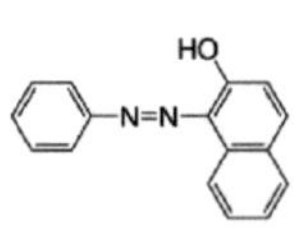 Sudan I 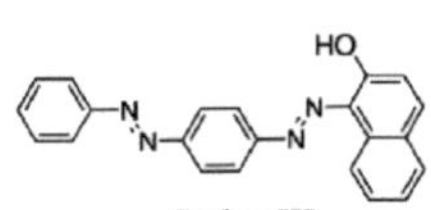 Sudan III 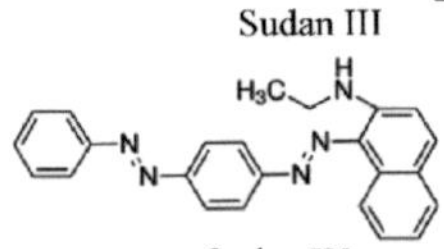 Sudan IV	Irreversible oxidation	[55]

Electrode	Dye/structure	Electrochemical behaviour	Reference
Glassy carbon electrode.	Reactive Brilliant Red x (RBR x-3b); Acid Red 6b (AR 6b); Reactive Yellow x-rg (RY x-rg); Reactive Orange x-gn (RO x-gn) 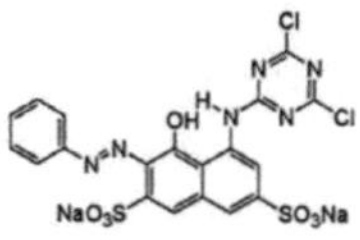 RBR x- 3b 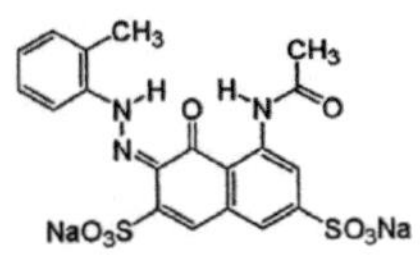 AR 6b RO x- gn 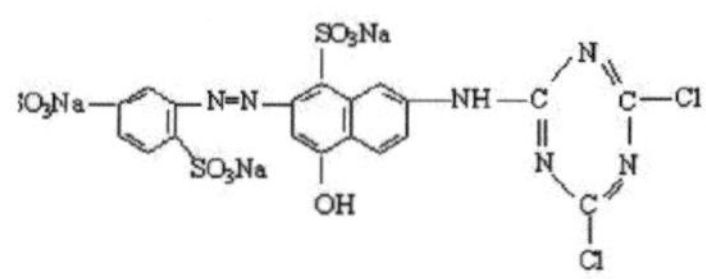 RY x- rg 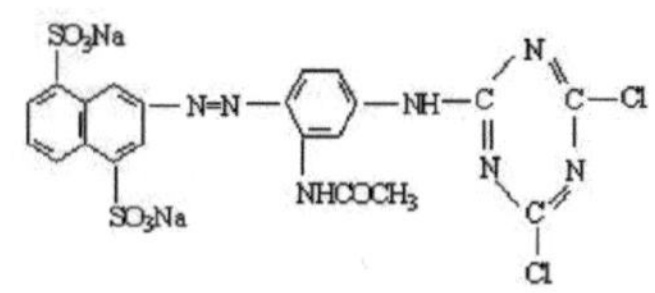	Irreversible oxidation	[59]
Graphite carbon electrodes.	Novacron Deep Red C-D; 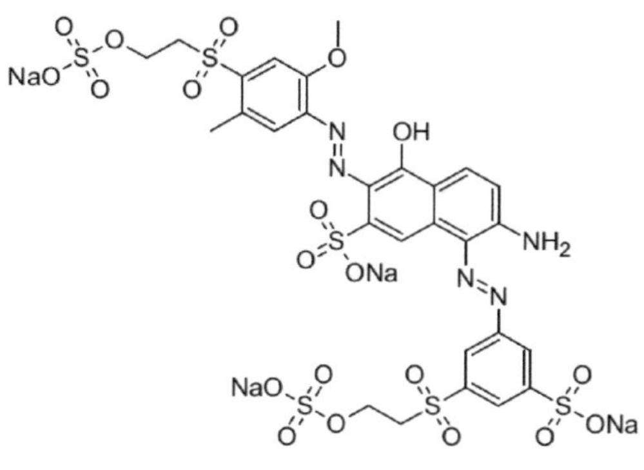 Novacron Orange C-RN 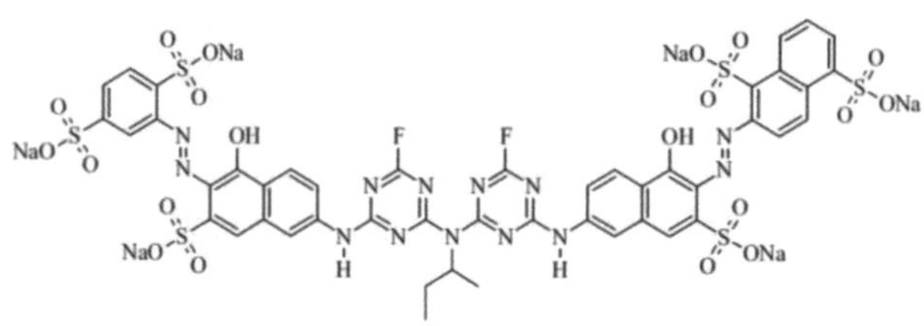	Reversible oxidation reduction	[60]

Table 1.
Electrochemical behaviour of some dyes investigated in electrochemistry.

3.2 Detection of metals using their reaction with dyes

The coordination complexes of metals with azo-ligands are used in several applications due to the interesting material properties synthetized. The metal complexation by dyes modify the photophysical and coloristic properties of dyes. The formation of complexes are influenced by several by parameters such as dye concentration, dye structure, pH, temperature, solvents and ionic strengths [61]. This reaction is due to the interactions by Van der Waals forces, hydrogen bonds and hydrophobic interactions [62]. In electrochemistry, the metals complexation with dyes has been investigated for the selective determination of metals trace by voltammetric techniques. For this, the oxidation/reduction current of dye is measured in the presence and the absence of metal ions. The decrease of the redox peaks caused by the formation of electro inactive complexes is function of the metals concentration and allows their determination. Thus, an electrochemical method for Cu(II) determination based on its reaction with indigo carmine (IC) in alkaline medium and differential pulse voltammetry performed at graphite electrode, was elaborated [51]. When Cu(II) ions are added to an alkaline solution of IC, the Cu2IC complex is formed [51]. This complex electro-inactive at the working potential trains the decrease of the oxidation peak current, which is depending on Cu(II) concentration. The detection limit was 4.74 μM [51]. The complexation studies between the indigo carmine food dye and mercury were carried out [63] and used for the determination of Hg (II) by electrochemical method. These studies show the suitability of voltammetric detection of metals trace using a decrease of oxidation/reduction current of dyes due to the formation of electro inactive metal -dye-complexes. However, there is very little work in literature concerning this approach. In general, these are electrodes modified by dyes which are used for the detection of metals in electrochemistry.

3.3 Dye modified electrodes for the determination of metal

The chemically modified electrodes have received an increasing attention in recent years in the fields of electroanalysis due to well recognized advantages in comparison with conventional electrodes [23]. The methods for modified electrode preparation are varied. The construction of dyes modified electrodes can be done by electrodeposition [64], by sol–gel method, by incorporation of dye into the carbon paste [65], by composite dye film or by the polymerization of dye on electrode surface. Thus, electrodes modified by chitosan -dye-enzyme composite film or copper complex dye (C.I. Direct Blue 200) film have been reported [64]. Studies have also been carried out on the polymerisation of dyes as modified agent. The poly-congo red (PCR) has been used to modified electrodes by electro-deposition [66, 67] or by polycondensation [5]. However, in voltammetric measurement, the polymerizations of congo red reported are generally carried out with the incorporation of poly congo into other components. These components can be polymers such as aniline [58] or nanoparticles [66]. We will notably encounter the synthesis of poly pyrole on glassy carbon in the presence of Congo red in view of the detection of dopamine [67], but also the use of poly Congo red in composition with nanoparticles such as CdS for the quantification of α-1-fetoprotein [66], several studies also relate to the use of Congo red in the presence of nanotube [59]. However, the use of dye as modifying reagent in electrochemical analysis is limited. Very little works about this topic have been reported in literature.

4. Conclusion

Dyes are widely used for industrial, printing, food, cosmetic and clinical purposes as well as in analytical chemistry. They play an important role in spectrophometry and electrochemistry analysis as complexing reagent for the detection of metal ions but they potential is still underestimated in electroanalysis.

It is well known that thousands tons of synthesis dyes are annually produced worldwide. Despite considerable work in recent years on the synthesis of new dyes which can be used reagent for the determination of trace amounts metals, the use of dye as modifying agents is still limited in electroanalysis. Further investigations are required to synthesis dyes in function of metal and use them to modified electrodes for electrochemical determination of metal ions.

5. Future outlook

The ill effects of metals on health and the environment are well documented, yet there is a lack of reliable, robust, cheap, and accurate sensors for monitoring toxics metals level in environment. Although much progress has been achieved in the last few decades in the field of chemically modified electrodes, techniques developed so far in electroanalysis have used very little the properties of dyes. This signifying that more research and development of new electrodes or new metal complexes for spectrophotometric analysis are required. The inconvenients of current sensors concern largely on their reproductibilty, sensitivity and their selectivity toward target metal. The cost associated with portable nature and speciation analysis are also a limiting factor.

Thus there is an urgent need to synthesis new dyes in function of target metal for the application to the electroanalysis of metal cation and in spectrophotometric analysis. Several investigations can therefore be carried out: synthesis of new dye as ligand for metal complexation; study of physicochemical characteristics of new synthesis ligands; study of their electrochemical behavior and complexation; characterization of metal dyes complexes for the electro and spectro analysis; polymerization of dye; thermodynamic study of the complexation of metal cations in poly-dye/poly film dye; study of chemically modified electrodes (ECM) by dye and their application to the detection of metal ions, etc.

Author details

Mariame Coulibaly
Ecole Normale Superieure d'Abidjan, Abidjan, Côte d'Ivoire

*Address all correspondence to: coulibaly.mariame@ensabj.ci; mamecoul2002@yahoo.fr

References

[1] Bauer C, Jacques P, kalt A. Photooxidation of an azo dye induced by visible light incident on the surface of Tio2. J. Photochem. Photobiol. A Chem. 2001; 140: 87-92. DOI: 10.1016/S1010-6030(01)00391-4

[2] Ganesh R, Boardman GD, Michelson D. Fate of azo dyes in sludges. Water Res. 1994; 28: 1367-1376. DOI:. org/10.1016/0043-1354(94)90303-4

[3] O'Neill C, Hawkes FR, Hawkes DL, N.D. Lourenco ND, Pinheiro HM, W. Delle W.. Colour in textile effluents – sources, measurement, discharge consents and simulation: a review. J. Chem. Technol. Biotechnol. 1999; 74: 10009-10018. DOI: org/10.1002/(SICI)1097-4660(199911)74:11<1009::AID-JCTB153> 3.0.CO;2-N.

[4] Pandey A, Sing, Iyengar (2007). Bacterial decolorization and degradation of azo dyes. Int. Biodeter. Biodegrad. 2007; 59: 73-84. DOI: org/10.1016/j.ibiod.2006.08.006

[5] Benkhaya S, Mrabet S, El Harfi A. Classifications, properties, recent synthesis and applications of azo dyes. Heliyon. 2020; 6: 32712. DOI: https://doi.org/10.1016/j.heliyon.2020.e03271

[6] Christie, R., 2001. Colour Chemistry. The Royal Society of Chemistry, Cambridge, United Kingdom.

[7] Luqman JR, Shahid-ul-Islam SA, Qazi PH, Faqeer M. Chemistry of Plant Dyes: Applications and Environmental Implications of Dyeing Processes. Current Environmental Engineering. 2017; 4: 103 – 120. DOI: 10.2174/2212717804666161216114949

[8] Zollinger H. Color chemistry: syntheses, properties and applications of organic dyes and pigments. 2.ed. New York: VCH; 1991. 496 p. DOI: hlib.ugent.be/catalog/rug01:000277149

[9] Lima EC, Royer B, Vaghetti JCP, Brasil JL. Adsorption of Cu(II) on Araucaria angustifolia wastes: Determination of the optimal conditions by statistic design of experiments. Journal of Hazardous Materials. 2007; 140: 211-220. DOI: https://doi.org/10.1016/j.jhazmat.2006.06.073

[10] Huo J, Guo Y, Meng S, Wang M, Wang Y, Bioinformatics and Biomedical Engineering (iCBBE), 2010 4th International Conference.

[11] Bazel Y, Tupys A, Ostapiuk Y, Tymoshuk O, Imrich J, Šandrejová J. A simple non-extractive green method for the spectrophotometric sequential injection determination of copper(II) with novel thiazolylazo dyes. RSC Adv.2018; **8:** 15940-15950 DOI: 10.1039/C8RA02039F

[12] Gupta V K, Ganjali, M R. Norouzi, P. Khani H. Nayak A, Agarwal S. Electrochemical Analysis of Some Toxic Metals by Ion–Selective Electrodes. Critical Reviews in Analytical Chemistry. 2011; 41: 282-313 DOI: https://doi.org/10.1080/10408347.2011.589773

[13] Aragay G, Merkoçi A. Nanomaterials application in electrochemical detection of heavy metals. Electrochim. Acta. 2012; 84: 49-61. DOI: 10.1016/j.electacta.2012.04.044

[14] Buica G O, Bucher C, Moutet J.-C, Royal G, Saint-Aman E, Ungureanu, E M. Voltammetric sensing of mercury and copper cations at poly(EDTA-like) film modified electrode. Electroanalysis. 2009;21, 77-86. DOI: 10.1002/elan.200804386

[15] Lee S, Bong S, Ha J, Kwak M, Park S, Piao Y. Electrochemical deposition of bismuth on activated graphene-nafion composite for anodic stripping voltammetric determination of trace heavy metals. Sensors and Actuators B:

Chemical. 2015; 215: 62-69 DOI: https://doi.org/10.1016/j.snb.2015.03.032.

[16] Wang Y, Xu H, Zhang J, Li G. Electrochemical Sensors for Clinic Analysis. Sensors. 2008; 8: 2043-2081. DOI: 10.3390/s8042043.

[17] Barbeira PJS, Mazo LH, Stradiotto NR. Determination of trace amounts of zinc, lead and copper in sugar cane spirits by anodic stripping voltammetry. Analyst. 1995; 120: 1647-1650. DOI: org/10.1039/AN9952001647

[18] Economou A, Fielden PR. Mercury film electrodes: developments, trends and potentialities for electroanalysis. Analyst. 2003; 128: 205-213. DOI: org/10.1039/B201130C

[19] Coulibaly M, El Rhazi M, Adraoui I. Determination of traces of Copper by anodic stripping voltammetry at a rotating carbon paste disk electrode modified with poly (1, 8-Diaminonaphtalene). Journal of Analytical Chemistry. 2009; 64: 632 – 636. DOI: https://doi.org/10.1134/S1061934809060161.

[20] Salih FE, Ouarzane A, El Rhazi M, Electrochemical detection of lead (II) at bismuth/Poly (1,8-diaminonaphthalene) modified carbon paste electrode. Arabian Journal of Chemistry. 2017; 10: 596-603 DOI: https://doi.org/10.1016/j.arabjc.2015.08.021.

[21] Svancara I, Prior C, Hocevar SB, Wang J. A decade with bismuth-based electrodes in electroanalysis. Electroanalysis. 2010. 22, 1405-1420. DOI: 10.1002/elan.200970017

[22] Guo Z, Feng F, Hou Y, Jaffrezic-Renault, N. Quantitative determination of zinc in milkvetch by anodic stripping voltammetry with bismuth film electrodes. Talanta.2005; 65: 1052-1055. DOI: 10.1016/j.talanta.2004.08.060

[23] Kalcher K, Kaufmann JM, Wang J. Švancara I, Vytřas K, Neuhold C, Yang Z. Sensors based on carbon paste in electrochemical analysis: A review with particular emphasis on the period 1990-1993. Electroanalysis. 1995; 7: 5-22. DOI: https://doi.org/10.1002/elan.1140070103

[24] He Y, Wang Z, Ma L, Zhou L, Jiang Y, Gao J. Synthesis of bismuth nanoparticle-loaded cobalt ferrite for electrochemical detection of heavy metal ions. RSC Adv. 2020; **10:** 27697-27705. DOI: 10.1039/D0RA02522D

[25] Kokkinos C, Economou A, Pournara A, Manos M, Spanopoulos I, Kanatzidis M, Tziotzi T, Petkov V, Margariti A, Oikonomopoulos P, Papaefstathiou GS. 3D-printed lab-in-a-syringe voltammetric cell based on a working electrode modified with a highly efficient Ca-MOF sorbent for the determination of Hg(II). Sensors and Actuators B: Chemical. 2020; 321: 128508. DOI: https://doi.org/10.1016/j.snb.2020.128508.

[26] Wang Z, Liu E, Zhao X. Glassy carbon electrode modified by conductive polyaniline coating for determination of trace lead and cadmium ions in acetate buffer solution, Thin Solid Films. 2011; 519: 5285-5289, DOI:https://doi.org/10.1016/j.tsf.2011.01.176.

[27] Hassan KM, Gaber SE, Altahan MF, Azzem MA. Single and simultaneous voltammetric sensing of lead(II), cadmium(II) and zinc(II) using a bimetallic Hg-Bi supported on poly(1,2-diaminoanthraquinone)/glassy carbon modified electrode. Sensing and Bio-Sensing Research.2020; 29:100369. DOI: https://doi.org/10.1016/j.sbsr.2020.100369.

[28] Salih FE, Achiou B, Ouammou M, Bennazha J, Ouarzane A, Younssi SA, El Rhazi M, Electrochemical sensor based on low silica X zeolite modified carbon paste for carbaryl determination. Journal of Advanced Research. 2017; 8: 669-676, DOI: https://doi.org/10.1016/j.jare.2017.08.002.

[29] Ravichandran K. & Baldwin R. P. 1981. Chemically modified carbon paste electrodes. Journal of Electroanalytical Chemistry and Interfacial Electro-chemistry,vol. 126,no1-3, p.293-300.

[30] El Mhammedi M. A., Bakasse M., Najih R. & Chtaini A. 2009. A carbon paste electrode modified with kaolin for the detection of diquat. Applied Clay Science, vol.43, no 1, p.130-134

[31] Estrada-Aldrete J, Hernández-López JM, García-León AM, Peralta-HernándezJM,Cerino-CórdovaFJ. Electroanalytical determination of heavy metals in aqueous solutions by using a carbon paste electrode modified with spent coffee grounds. Journal of Electroanalytical Chemistry.2020; 857: 113663. DOI: https://doi.org/10.1016/j.jelechem.2019.113663.

[32] Coulibaly M, Ghanjaoui M E, Gonzalvez A, DeLaguardia M, El Rhazi M. Determination of selenium (IV) in pharmaceutical products by differential pulse voltammetry and inductively coupled plasma optical emission spectroscopy, Arabian. J. Chem. 2008; 1: 307 – 317.

[33] Chen SM, Liu JW, Thangamuthu R. Preparation, Characterization and Electrocatalytic Studies on Copper Complex Dye Film Modified Electrodes. Electroanalysis.2007; 19: 1429-1436. DOI: https://doi.org/10.1002/elan.200703875

[34] Manisankar P, Gomathi A. Electrocatalytic Reduction of Dioxygen at the Surface of Carbon Paste Electrodes Modified with 9,10-Anthraquinone Derivatives and Dyes Electroanalysis. 2005; 17: 1051-1057. DOI: https://doi.org/10.1002/elan.200403213

[35] Miao X, Yuan R, Chai Y, Shi Y, Yuan Y. Electrochemical immunoassay for α-1-fetoprotein based on CdS nanoparticles and Thionine bilayer films modified glass carbon electrode. Biochemical Engineering Journal. 2008; 38: 9-15. DOI: https://doi.org/10.1016/j.bej.2007.05.015

[36] Tavallali H, Shaabanpur E, Vahdati. P. A highly selective optode for determination of Hg (II) by a modified immobilization of indigo carmine on a triacetylcellulose membrane. Spectrochimica Acta Part A. 2012; 89: 216-221.DOI: 10.1016/j.saa.2011.12.055

[37] Zanoni TB, Cardoso AA, Zanoni MVB, Ferreira AAP. Exploratory study on sequestration of some essential metals by indigo carmine food dye. Brazilian Journal of Pharmaceutical Sciences2010; 46:723-730. DOI: https://doi.org/10.1590/S1984-82502010000400014

[38] Akram D, Elhaty IA, AlNeyadi SS. Synthesis and Antibacterial Activity of Rhodanine-Based Azo Dyes and Their Use as Spectrophotometric Chemosensor for $Fe3+$ Ions. Chemosensors. 2020; 8:1-13. DOI:10.3390/chemosensors8010016.

[39] Gürses A, Açıkyıldız M, Günes K, Gürses MS. Classification of dye and pigments, in: Dyes and Pigments, Springer, 2016, pp. 31-45.DOI: https://doi.org/10.1007/978-3-319-33892-7_3

[40] Chakraborty JN. Metal-complex dyes. Handbook of Textile and Industrial Dyeing Principles, Processes and Types of Dyes in Woodhead Publishing Series in Textiles 2011; 1: 446-465. DOI: https://doi.org/10.1533/9780857093974.2.446.

[41] Chavan RB. Environmentally friendly dyes. Handbook of Textile and Industrial Dyeing Principles, Processes and Types of Dyes in Woodhead Publishing Series in Textiles 2011; 1: 515-561. DOI: https://doi.org/10.1533/9780857093974.2.446

[42] P. Gregory, in Comprehensive Coordination Chemistry II, 2003

[43] James, A. K. (1997). Riegel's Handbook of Industrial Chemistry. (9th

Ed).CBS Publishers & Distributors PVT LTD. New Delhi, India

[44] Kilincarslan R, Erdem E, Kocaokutgen H. Synthesis and spectral characterization of some new azo dyes and their metal complexes. Trans Met Chem. 2007; 32:102-106 doi:10.1007/s11243-006-0134-x

[45] Mahapatra BB, Ajith Kumar NP, Bhoi PK. Polymetallic complexes. Part-XXX. complexes of cobalt-, nickel-, copper-, zinc-, cadmium- and mercury(II) with doubly-tridentate chelating azo-dye ligand. Journal of the Indian Chemical Society. 1990; 67: 800-802. DOI: https://www.osti.gov/etdeweb/biblio/5053854

[46] Guozhen C, Ultraviolet-visible Spectroscopy (Part 1)1983. Atomic Energy Press, Beijing, 1983.

[47] Ashok K. Sharma S. Spectrophotometric Trace Determination of Iron in Food, Milk, and Tea Samples using a New Bis-azo Dye as Analytical Reagent. Food Anal. Methods. 2009; 2:221 – 225. DOI 10.1007/s12161-008-9054-z

[48] Bonishko OS, Vrublevska TY, Zvir O Z, Dobryanska OP. Spectrophotometric of osmium (IV)ions in intermetallic compounds. Materials Science. 2008; 44: 248-253. DOI:https://doi.org/10.1007/s11003-008-9059-1

[49] Rydchuk M, Vrublevska T, Korkuna O, Volchak M. Application of Orange G as a Complexing Reagent in Spectrophotometric Determination of Osmium(IV). Chem. Anal. (Warsaw).2009; 54: 1051- 1063. DOI: http://www.chem.uw.edu.pl/chemanal/

[50] Chailapakula O, Wonsawat W, Siangprohb W, Grudpanc K, Zhaod Y, Zhud Z. Analysis of sudan I, sudan II, sudan III, and sudan IV in food by HPLC with electrochemical detection: Comparison of glassy carbon electrode with carbon nanotube-ionic liquid gel modified electrode. www.ift.org (newBody.shtml,in.). news_bin. news_bin.

[51] Coulibaly M, Mureşan LM, Popescu IC. Detection of Cu(II) using its reaction with indigo carmine and differential pulse voltammetry. Studia Chemia Journal. 2012; 3: 65-72. http://chem.ubbcluj.ro/~studiachemia/issues/chemia2006_2015/Chemia2012_3.pdf

[52] Tvorynska S, Bohdan, Barek J, Dubenska L. Electrochemical Behavior and Sensitive Methods of the Voltammetric Determination of Food Azo Dyes Amaranth and Allura Red AC on Amalgam Electrodes. Food Analytical Methods. 2018; 12: 409-421 DOI: https://doi.org/10.1007/s12161-018-1372-1

[53] Y. Zhang, L. Hu, X. Liu, B. Liu, K. Wu, Highly-sensitive and rapid detection of Ponceau 4R and Tartrazine in drinks using alumina microfibers-based electrochemical sensor, Food Chem. 166 (2015) 352-357, https://doi.org/10.1016/j.foodchem.2014.06.048.

[54] D. Sun, C. Xu, J. Long, T. Ge, Determination of Sunset Yellow using a carbon paste electrode modified with a nanostructured resorcinol-formaldehyde resin, Microchim. Acta 182 (2015) 2601-2606, https://doi.org/10.1007/ s00604-015-1646-x.

[55] Y. Tang, Y. Wang, G. Liu, D. Sun, Determination of Sunset Yellow and Tartrazine using silver and poly (L-cysteine) composite film modified glassy carbon electrode, Indian J. Chem. 55A (2016) 298-303

[56] Shetti NP, Malodea S, Malladi RS, Nargund SL, Shuklad SS, Aminabhavi TM. Electrochemical detection and degradation of textile dye Congo red at graphene oxide modified electrode. Microchemical Journal.

2019; 146: 387-392. DOI:https://doi.org/10.1016/j.microc.2019.01.033

[57] Silva MLS, Garcia MBQ, Lima JL, Barrado E. Voltammetric determination of food colorants using a polyallylamine modified tubular electrode in a multicommutated flow system. Talanta. 2007; 72: 282-288. DOI: https://doi.org/10.1016/j.talanta.2006.10.032.

[58] Shetti NP, Nayak DS, Malode SJ. Electrochemical behavior of azo food dye at nanoclay modified carbon electrode-a nanomolar determination. Vacuum. 2018; 155: 524-530. DOI: https://doi.org/10.1016/j.vacuum.2018.06.050

[59] Yu J, Jia J, Ma Z. Comparison of Electrochemical Behavior of Hydroxyl-substituted and Nonhydroxyl-substituted Azo Dyes at a Glassy Carbon Electrode. Journal of chinese Chemical Society. 2004; 51: 319-1324. DOI; https://doi.org/10.1002/jccs.200400191

[60] Kariyajjanavar P, Jogttappa N, Nayaka YA. Studies on degradation of reactive textile dyes solution by electrochemical method. J Hazard Mater. 2011; 190:952-61 DOI: 10.1016/j.jhazmat.2011.04.032

[61] Goftar MK, Moradi K, Kor NM. Spectroscopic studies on aggregation phenomena of dyes. European Journal of Experimental Biology. 2014; 4: 72-81.

[62] Radulescu-Grad ME, Muntean SG, Todea A., Verdes O. Synthesis and characterization of new metal complex dye. Chemical Bulletin of Politehnica University of Timisoara. 2015; 60: 37-40.

[63] Seiny NR, Coulibaly M, Yao AN, Bamba D, Zoro EG. An Electrochemical Method for the Determination of Trace Mercury (II) by Formation of Complexes with Indigo Carmine Food Dye and Its Analytical Application. Int. J. Electrochem. Sci.2016; 11: 5342 – 5350, DOI: 10.20964/2016.06.61.

[64] Yang C, Xu Y, Hu C, Shengshui Hu S. Voltammetric Detection of Ofloxacin in Human Urine at a Congo Red Functionalized Water-Soluble Carbon Nanotube Film Electrode.Electroanalysis. 2008; 20: 144 – 149. DOI: https://doi.org/10.1002/elan.200704027

[65] LeeTS, Yang C. Synthesis of Congo Red linked with alkyl amide polymer and its optical ion-sensing property. Polymer Bulletin. 1999; 42; 655-660. DOI: https://doi.org/10.1007/s002890050515

[66] Pavan FA, Ribeiro ES, Gushikem Y. Congo Red Immobilized on a Silica/Aniline Xerogel: Preparation and Application as an Amperometric Sensor for Ascorbic Acid. Electroanalysis. 2005;17: 625-629 DOI: https://doi.org/10.1002/elan.200403132

[67] Shahrokhian S, Zare-Mehrjardi HR, Khajehsharifi H. Modification of carbon paste with congo red supported on multi-walled carbon nanotube for voltammetric determination of uric acid in the presence of ascorbic aci. J Solid State Electrochem. 2009; 13:1567-1575. DOI: https://doi.org/10.1007/s10008-008-0733-x

Chapter 8

Structure and Properties of Dyes and Pigments

Ashok Kumar, Utkarsh Dixit, Kaman Singh, Satya Prakash Gupta and Mirza S. Jamal Beg

Abstract

Colour is one of the elements of nature that makes human life more aesthetic and fascinating in the world. Plants, animals, and minerals have been used as primary sources for colourants, dyes or pigments since ancient times. In our daily life, we know about many substances which have specific colours. These are the substances which are used as colourants i.e.; colour imparting species. Both dyes and pigments are coloured as they absorb only some wavelength of visible light. Their structures have Aryl rings that have delocalized electron systems. These structures are said to be responsible for the absorption of electromagnetic radiation that has varying wavelengths, based upon the energy of the electron clouds. Dyes are coloured organic compounds that are used to impart colour to various substrates, including paper, leather, fur, hair, drugs, cosmetics, waxes, greases, plastics and textile materials. A Dye is a coloured compound due to the presence of chromophore and its fixed property to the acid or basic groups such as OH, SO_3H, NH_2, NR_2, etc. The polar auxochrome makes the dye water-soluble and binds the dye to the fabric by interaction with the oppositely charged groups of the fabric structure. Pigments are organic and inorganic compounds which are practically insoluble in medium in which they are incorporated. Dyes and pigments are the most important colourants used to add colour or to change the colour of something. They are widely used in the textile, pharmaceutical, food, cosmetics, plastics, paint, ink, photographic and paper industries. This chapter is devoted to the structure and properties of dyes and pigments.

Keywords: Structure, Colourants, Chromophore, Auxochrome

1. Introduction

Colour provides a significant glimpse of our world. Everyday materials we tend to use different kinds of materials like - textiles, paints, plastics, paper, and foodstuffs. Colours make them most appealing. In summer there is a wild burst of colourful flowers and new leaves of various shades of green on trees [1]. However, in contrast, autumn makes the beautiful impression with green leaves turn to brilliant shades of yellow, orange, and red. Colour derives from the spectrum of light interacting in the eye with the spectral sensitivities the light receptors [1, 2].

A dye is nothing but a coloured substance that has an affinity to the substrate to which it is being applied. The dye is applied in an aqueous solution and needs a mordant to boost the fastness of the dye on the textile fibre. The pigment may be a material that modifies the colour of mirrored or transmitted light as the result of wavelength-selective absorption. Pigments are used for colouring paint, ink, plastic, fabric, cosmetics, food and other materials.

Both dyes and pigments appear to be coloured as a result of absorption of some wavelengths of light more than others. However, there are some basic differences between the dyes and pigments

- The major difference between the dyes and the pigments is that particle size of pigments is much higher as compared to dyes. Due to this small particle size dyes are not UV stable whereas pigments are UV stable.
- Dyes after dissolving in liquid are absorbed on the material while pigments make a suspension with a liquid that bonds with the material surface.
- Dyes are generally soluble in water while pigments are almost insoluble in water.
- Most of the dyes are organic while most of the pigments are inorganic
- Dyes are available in large number in the market while the number of pigments is very less.
- Dyes impart colour by selective adsorption while pigments adsorb colour either by selective adsorption or by scattering of light.
- Dyes are combustible while pigments are non-combustible.
- Dyes have a short lifetime in comparison to pigments.

Over the years, man has used colouring matters, which are known as dyes and pigments, for their aesthetic qualities and used them to embellish various articles and the world in which he lived. Indigo, the oldest known dye, was discovered in India; Tyrian purple (or Royal Purple) was discovered in the ancient city of Tyre; Alizarin was discovered among the Turks; and Cochineal was discovered among European and Mexican dyers [3]. Indigenous dye-yielding plants have been discovered in almost every area of the world. The first synthetic dyes were found in the early twentieth century. Parenteral administration was not formulated until the 1930s: methylene blues and methyl violet, for example, were used to treat leprosy and filariasis, respectively [4]. Following World War II, the use of intravenous dyes for medicinal purposes decreased rapidly. Just a few dyes, such as patent blue V or fluorescein, are still used as diagnostic drugs today [5].

The textile industry now uses more synthetic dyes. Coal tar and petroleum-based intermediates are the two main sources of these chemicals. Powders, granules, pastes, and liquid dispersions are all available [6, 7]. Active ingredient concentrations usually vary from 20 to 80 percent. The textile dye segment is distinguished by the introduction of new dyes. These new dyes are produced on a regular basis to meet the demands of new technologies, new types of fabrics, detergents, and developments in dyeing machinery, as well as to address the significant environmental issues posed by some existing dyes [8–10]. With the rapid shift in the textile industry's product profile, from high-cost cotton textiles to durable and flexible synthetic fibres, the pattern of dye use is also shifting rapidly. Polyesters now account for the majority of dye use. Disperse dyes, which are used in Polyesters, are expected to expand at a faster pace as a result. Textile dyestuffs may be grouped into the following groups for better comprehension if general dye chemistry is used as one of the classification criteria as acid dyes, direct dyes, azoic dyes, disperse dyes, sulphur dyes, reactive dyes, basic dyes, oxidation dyes, mordant dyes (chrome dyes) and vat dyes [11–15].

2. Historical background of dyes and pigments

In the ancient age usually used all the dyes were natural. Some of the natural dyes used in ancient age were alizarin and indigo. Indigo is probably the oldest known dye obtained from the leaves of dyers woad herb *Isatis tinctoria*, and the indigo plant *Indigofera tinctoria* [16]. Early dyes were obtained from animal, vegetable or mineral sources, with no to very little processing. The first synthetic dye, mauve, was discovered serendipitously by William Henry Perkin in 1856. The discovery of mauveine started a surge in synthetic dyes and inorganic chemistry in general.

Artists invented the first pigments—a combination of soil, animal fat, burnt charcoal, and chalk—as early as 40,000 years ago, creating a basic palette of five colours: red, yellow, brown, black, and white. In the early age man used earth pigments on cave walls such as yellow earth (ochre), red earth (ochre) and white chalk. Ochres are probably the oldest known pigments, which are coloured clays found as soft deposits within the earth [17–21].

2.1 Reason for the colour of a dye

Many theories have been given to correlate the colour of dyes with their molecular structure. In 1876, Otto Witt, a German Chemist observed that the colour of a dye is due the certain groups containing multiple bonds known as chromophores. Some examples of the chromophores are nitro group(-NO_2), nitroso group(-NO), carbonyl group(-CO-), ethylenic bond(-C=C), acetylenic bond etc. As the number of chromophores increases for a dye, the colour of the dye also deepens [17–21].

He also observed that not only the chromophores are responsible for the deepening of colour but also there are certain groups which itself does not act as the chromophore but the presence of which deepens the colour of the dye. These groups are known as Auxochromes. Some examples of the auxochromes are -OH, -NH_2, -NHR, -NR_2, X (Cl, Br or I), COOH. 1,3-Dinitronapthalene (**Figure 1**) is pale yellow but the dye Martius Yellow (2,4-Dinitro-1-naphthol) is orange-red (**Figure 2**) [17–21]. Here group -OH is acting as the autochrome as the presence of it has deepened the colour of 1,3-Dinitronapthalene.

To explain the origin of colour, Valence Bond Theory (VBT) and Molecular Orbital Theory (MOT) are proposed in modern time. The vital difference between the VBT and MOT is that in VBT electrons are treated in pairs while in MOT electrons are treated singly [17–22].

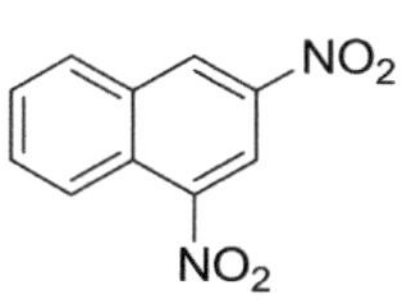

Figure 1.
1, 3-Dinitronaphthalene.

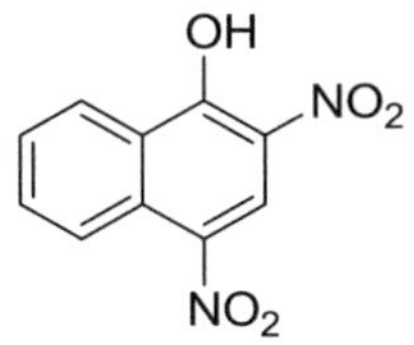

Figure 2.
2,4-Dinitro-1 naphthol.

I. **Valence bond theory (VBT):** According to VBT, in the ground state the electron pairs of a molecule are in a state of oscillation and absorb a photon of appropriate energy and get excited when placed in the path of a beam of light. The wavelength of a photon of light adsorbed depends upon the energy difference between the ground state and the excited state.

II. **Molecular orbital theory (MOT):** According to MOT, whenever a molecule absorbs a photon of light, one electron is transferred from bonding (non-bonding) orbital to an anti-bonding orbital. Based on different type of electron present in a molecule, different types of electronic transitions are possible.

3. Nomenclature

The chemical names of dyes are very complicated hence trade names are more popular instead of their chemical names. From the ancient time humans have tried to extract the dyes from plants and other natural sources to colour their clothes and other belongings, such dyes are known as Natural Dyes. Indigo and Alizarin are two examples of natural dyes. As the natural dyes have very few colours and shades so now a day's most of the used dyes are synthetic dyes having several colours and shades. Almost all synthetic dyes are aromatic and obtained from coal-tar [22]. Hence synthetic dyes are also called coal-tar dyes. Dyes can be classified by the following ways;

I. **Based on their constitution:** In this classification dyes are classified based on the functional group to which the dyes owe their colour. Some examples are *azo dyes, nitro dyes, nitroso dyes, triphenylmethane dyes, indigoid dyes, phthalein dyes, acridine dyes etc.*

II. **Based on their application:** The colouring of dye on a particular fibre depends on the nature of both the dye and the fibre. A dye molecule can be attached to fibre by following methods;

- Covalent Bonds
- Hydrogen Bonds
- Ionic Bonds
- Van Der Waals Forces

Based on application dyes can be categorised as follows;

I. **Acid dyes:** The sodium salt of azo dyes containing sulphonic acid ($-SO_3H$) and carboxylic acid($-COOH$) groups are called acid dyes. To colour the fabric from these dyes the acidic solution of these dyes is used. These can be used to colour wool, silk, nylon and polyurethane fibres. The affinity of acid dyes for nylon is very high due to the higher protonation of the free amino group present in polycaprolactam fibres. Orange-I (**Figure 3**), orange-II, methyl orange and congo red are some examples of it [22, 23].

The dye orange I and orange II can be obtained by coupling diazotised sulphonic acid with α and β- Naphthol respectively.

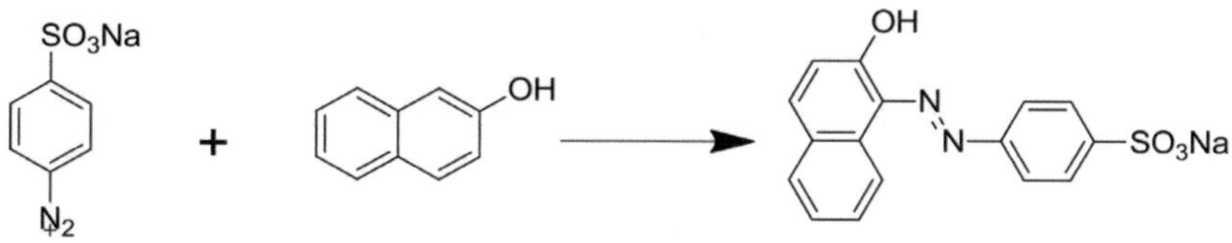

Diazotisedsulphonic β-Naphthol Orange-I acid sodium salt

II. **Basic dyes:** Those dyes which are salts of coloured bases containing amino or dialkylamino group as autochrome is known as Basic dyes. These include triphenylmethane and azo dyes. Modified nylons and polyesters can be dyed with the help of these dyes [22, 23]. Some examples of basic dyes are aniline yellow (**Figure 4**), butter yellow (**Figure 5**) and chrysodine G (**Figure 6**).

III. **Direct or substantive dyes:** These are water-soluble dyes hence they can be applied to the fabric directly [22, 23]. Congo red and Martius yellow (**Figure** 7) are two examples of these dyes.

IV. **Disperse dyes:** These are water-insoluble dyes and applied to the fabric in the form of a dispersion in presence of some stabilising agent such as phenol, cresol or benzoic acid. The two examples of disperse dyes are Celliton fast pink B and Celliton fast blue B [22, 23].

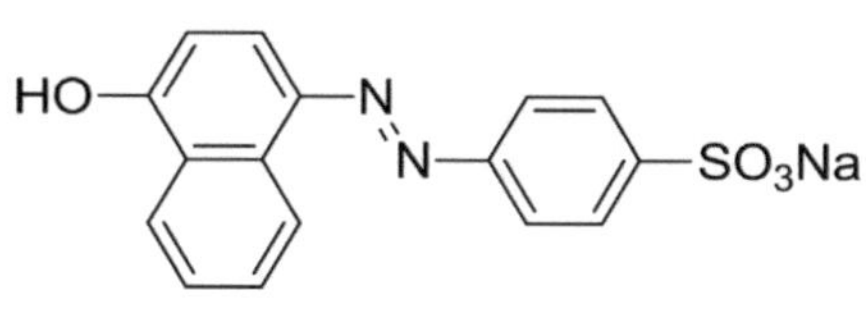

Figure 3.
Orange-I.

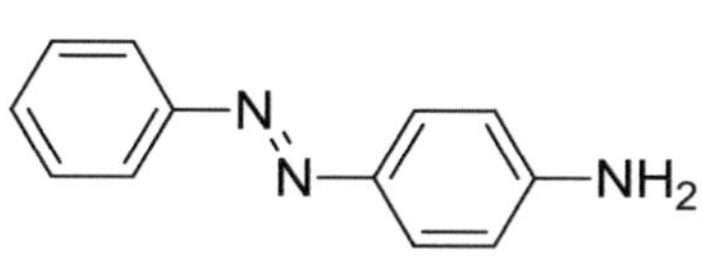

Figure 4.
Aniline yellow.

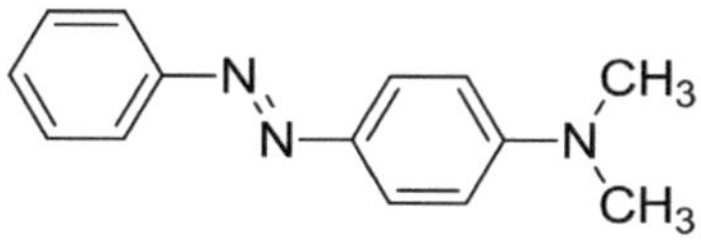

Figure 5.
Butter yellow.

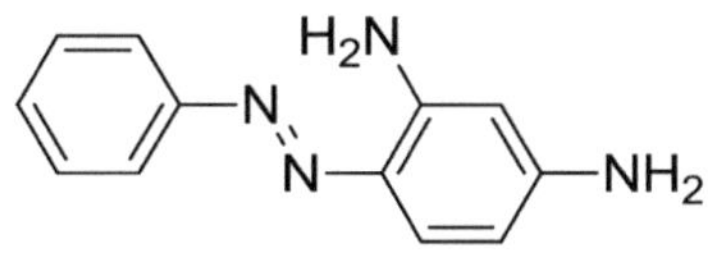

Figure 6.
Chrysodine G.

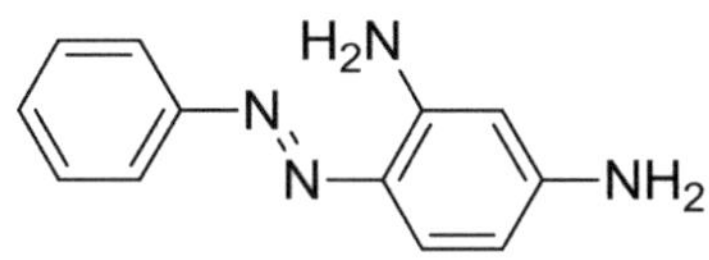

Figure 7.
Martius yellow.

V. **Fibre reactive dyes:** These dyes contain a reactive group which combines directly with the hydroxyl or amino group of the fibre. As a permanent bond formed between the fibre and dye in this case so the colour of the dyed fabric is very fast and has a long life [22, 23].

VI. **Ingrain dyes or insoluble azo dyes:** These constitute about 60% of total dyes used. These dyes are obtained by the coupling of phenols, naphthols, arylamines, aminophenols. These dyes are adsorbed only on the surface of the fabric, colouring by this dye is not so fast. These can be used to dye cellulose, silk, nylon and leather [22, 23]. Para red (**Figure 8**) is an example of such a class of dye. Para red (**Figure 8**) dye can be prepared as follow:

These dyes are also used in foodstuffs, cosmetics, drugs and as an indicator in chemical analysis.

VII. **Vat dyes:** Vat dyes are insoluble in water so they cannot be used directly for dyeing. They are first reduced to soluble colourless form (leuco form) with a reducing agent such as an alkaline solution of sodium hydrosulphite. Under these conditions, the leucoform develops an affinity for cellulose fibres [22, 23]. Hence these dyes are mainly used to dye cotton fibre. The example of vat dyes is indigo (**Figure 9**).

VIII. **Mordant dyes:** Those dyes which do not bind directly but require a mordant to dye the fabric directly come under this category. The mordant act as the binding agent between the fibre and the dye.

Metal ions are used as mordants for the acid dyes while tannic acid is used as the mordant for basic dyes. The desired fabric is first soaked in the suitable metal salt and then this soaked fibre is dipped in the solution of dye when insoluble coloured complexes formed on the fabric. These insoluble coloured complexes are called

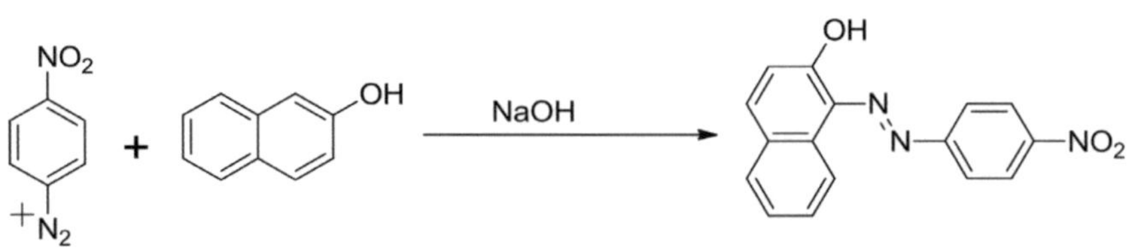

Figure 8.
Para red.

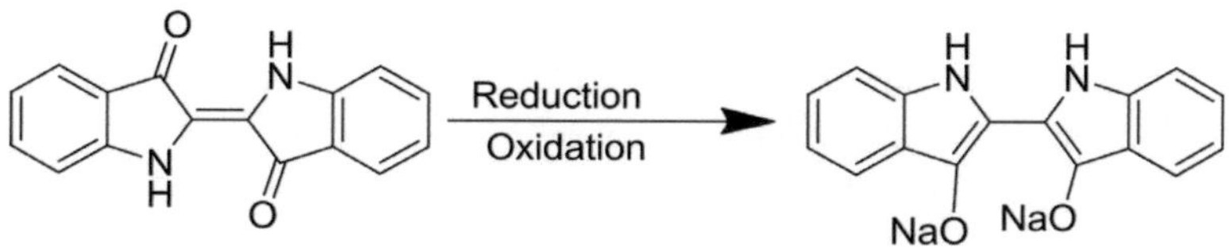

Figure 9.
Indigo or Indigotin (Leuco form).

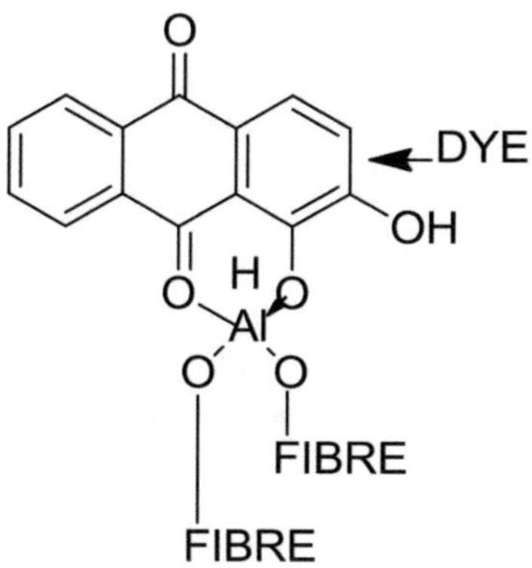

Figure 10.
Alizarin-aluminium fibre complex (rose red lake).

lakes [22, 23]. So, the metal ions first get attached to the fabric and then the dye molecules are linked to the metal ion through covalent and or coordinate bond (**Figure 10**).

4. Classification of dyes based on their constitution

4.1 Azo dyes

These dyes constitute the largest part of the synthetic dyes. The chromophore of the azo dyes is aromatic system joined to the azo group and auxochromes are NH_2, NR_2, OH.

Azo dyes are classified as the number of azo group in the molecule such as monoazo, diazo and triazole etc. The characteristic of two important azo dyes methyl orange (**Figure 11**) and congo red can be described as follows:

4.1.1 Methyl orange

It is obtained by coupling of N, N-dimethylaniline with diazotised sulphanilic acid.

Properties: Methyl orange is a colouring dye for the wool and silk but its colour fades on the exposure to the light and washing. Usually, it is not used as a dye but

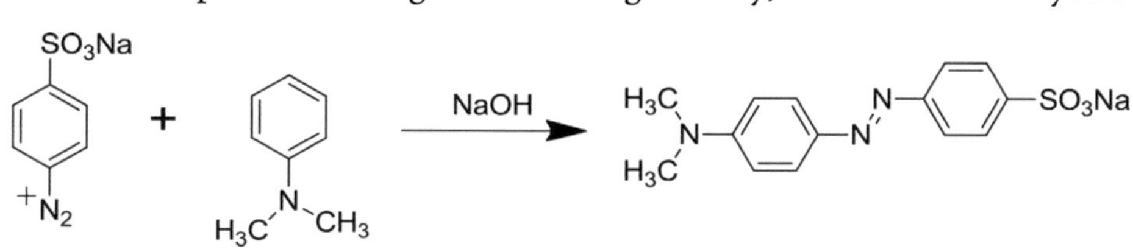

Figure 11.
Methyl Orange.

used as an indicator in the acid–base titrations. The pH range of methyl orange is 3.1–4.4. It is yellow in basic solutions (above pH 4.4) while red in acidic solutions (below pH 3.1) [24–26]. The colour change takes place because of the change in the structure of ions in acidic and basic medium. In acidic medium, the ion contains p-quinonoid chromophore while in basic medium ion contains azo chromophore (**Figure 12**).

4.1.2 Congo red

Congo red (**Figure 13**) is the example of diazo dye as it contains two azo groups. It is obtained by the coupling of tetrazotised benzidine with two molecules of napthionic acid (4-aminonapthalene-1-sulphonic acid).

Properties: it is a direct dye and sodium salt of this dye gives red colour on the applied cotton. As on addition of acid, its colour changes so it is also not used as a dye generally. Mostly it is used as an indicator. It is blue in acidic solution (below pH 3) and red in solutions (above pH 3). The change in colour from red to blue in the acidic solution is due to the resonance among charged canonical structures.

4.2 Triphenylmethane dyes

These are the derivatives of -NH_2, -NR_2, and -OH groups in the rings. The compound obtained by this method are generally colourless and called leuco bases. On oxidation, these are converted to the corresponding colourless tertiary alcohols called carbinol bases. These carbinol bases readily change from the colourless benzenoid form to the coloured quinonoid forms in the presence of acids due to salt formation, which is known to be dye [26–30]. The structure and properties of two important triphenylmethane dyes can be discussed as follows;

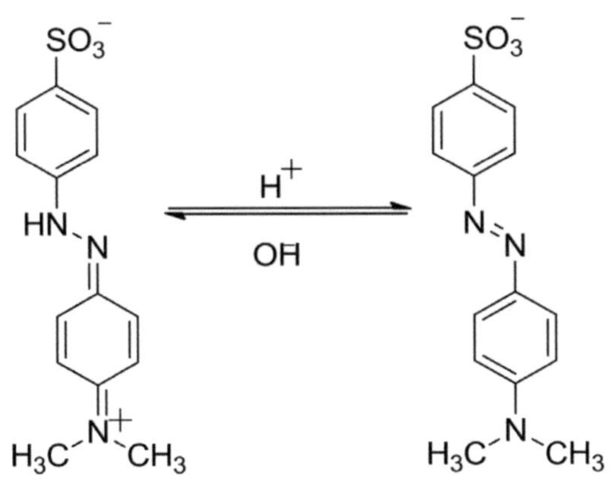

Figure 12.
Red (acidic medium) yellow (basic medium).

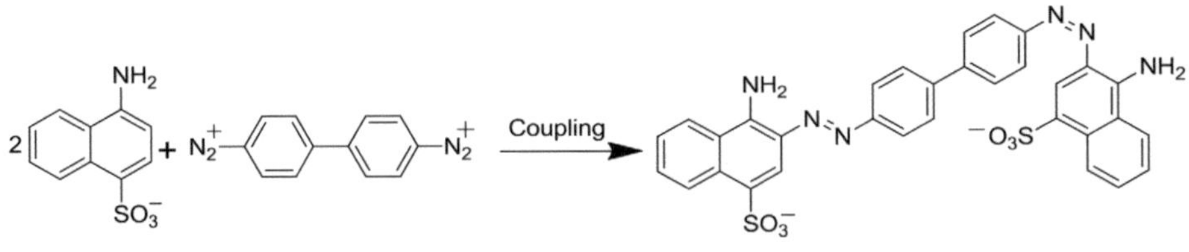

Figure 13.
Congo red.

4.2.1 Malachite green

It is used for dying wool and silk directly and cotton mordanted with tannin. The colour of it fades slowly on the addition of acid and base (**Figure 14**).

4.2.2 Crystal violet

A weakly acidic solution of it is purple, a strongly acidic solution is green and still more acidic solution is yellow (**Figure 15**).

4.3 Phthalein dyes

4.3.1 Phenolphthalein

It is insoluble in water but dissolves in alkalies to form a deep red solution. When an excess of strong alkali is added, the solution of phenolphthalein becomes colourless again. Because of the colour change, it is used as an indicator in acid–base titrations. It is also a powerful laxative (**Figure 16**).

4.3.2 Fluorescein

It is a xanthan derivative. In properties, it more closely resembles phthalein dyes. It is a red powder insoluble in water. It dissolves in alkalies to give a reddish-brown solution which, on dilution, gives a strong yellowish-green fluorescence (**Figure 17**).

4.3.3 Alizarin

It is one of the most important anthraquinone dye. It occurs in madder root in form of its glucoside called, ruberthyric acid.

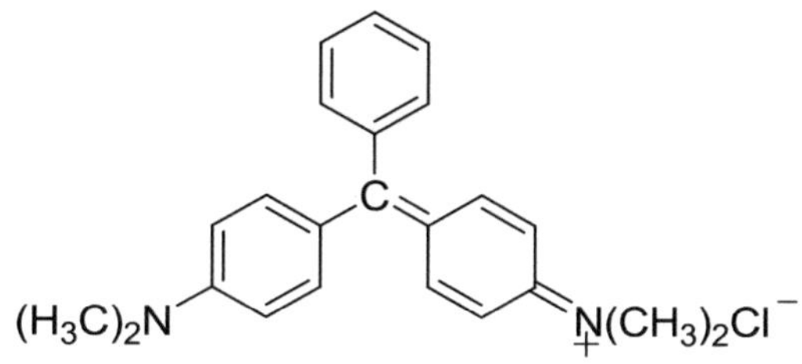

Figure 14.
Malachite green.

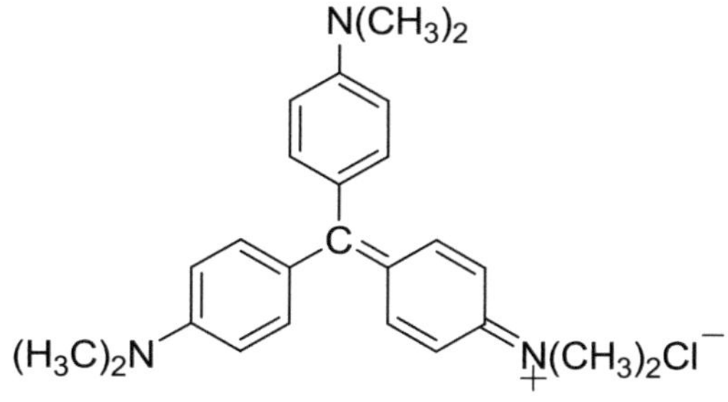

Figure 15.
Crystal violet.

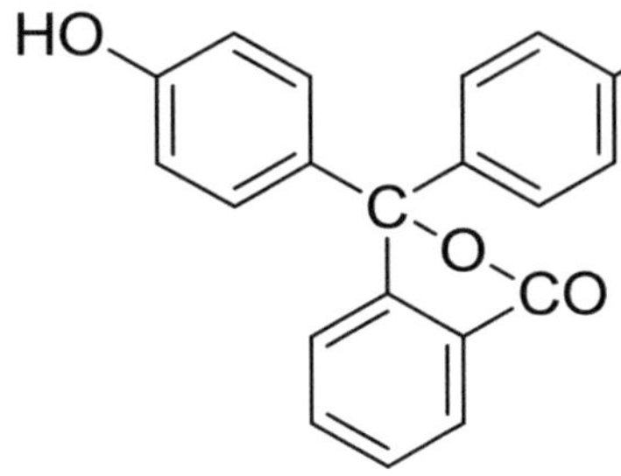

Figure 16.
Phenolphthalein.

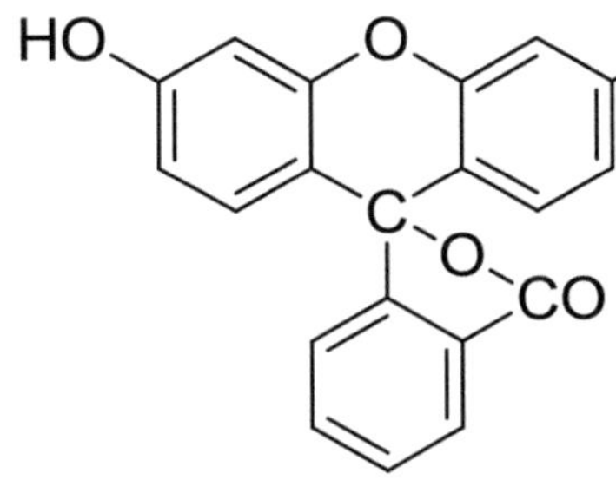

Figure 17.
Fluorescein.

On reduction with zinc dust, it gives anthracene. This implies that alizarin is a derivative of anthracene. It forms ruby red crystals, insoluble in water and alcohol but dissolves in alkalies to form purple solution, sublimes on heating.

It is a mordant dye, and the colour of the lake depends on the metal used. Aluminium gives a red lake, ferric salts give violet-black while chromium salts form a brown-violet lake. It is also used as a purgative (**Figure 18**).

4.3.4 Indigo

It is the oldest vat dye known. India is the birthplace of Indigo (**Figure 19**). A fusion of indigo at a low temperature produces anthranilic acid.

It is a dark blue powder, with coppered lustre, m.p. 663 K. It is insoluble in water and most organic solvents. It is widely used for dyeing cotton. The quality of colour is excellent and is stable to light, washing etc. It is also used in printing inks [26–30].

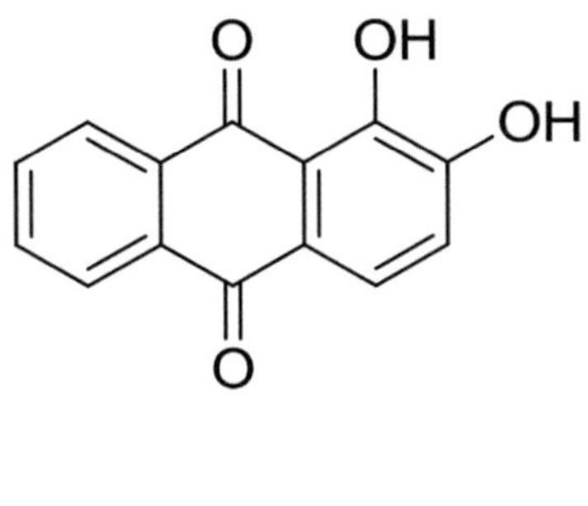

Figure 18.
Alizarin.

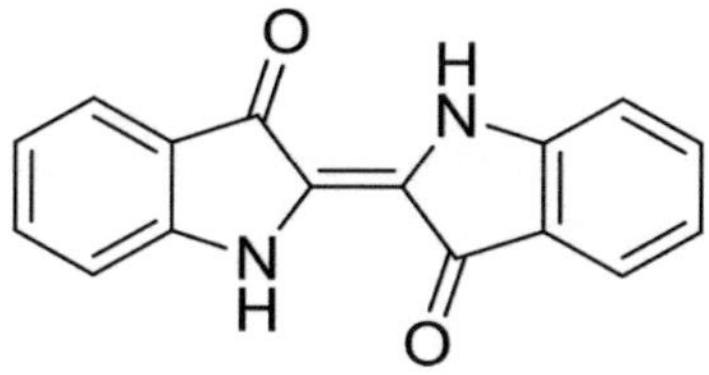

Figure 19.
Indigo.

5. Pigments

Those organic and inorganic substances which are widely used as surface coatings, employed in ink, plastic, rubber etc. to impart colour. A large number of pigments are used for commercial manufacture of paints.

5.1 History of pigments

Pigments are believed to be 3.5 lakh to 4 lakh years old. They have been reported in a cave at Twin Rivers, near Lusaka, Zambia. Blue pigment was derived from lapis lazuli. Pigments based on minerals and clays often bear the name of the city or region where they were originally mined. Synthetic pigments are believed to be introduced in early second millennium BCE. White lead $2PbCO_3$. $Pb(OH)_2$ and blue frit (Egyptian Blue) also known as calcium copper sulphate $CaCuSi_4O_{10}$ are the two examples of early synthetic pigments [17–30].

5.2 Classification of pigments

Pigments are mainly classified into two types:

1. White pigments

2. Coloured pigments

However, they may differentiate broadly into two category organic & inorganic pigments (**Table 1**) [23–26].

S.No.	Organic pigments	Inorganic pigments
1.	Pthalocyanines, Quinacidrones, Dioxazenes, Napthols, and other pigments made in the lab are examples	Examples include phthalocyanines, quinacridones, dioxazenes, napthols, and other lab-made pigments
2.	Plants and plant products are the primary source of organic pigments	Topical cosmetics, as well as dirt, are used to make inorganic pigments
3.	Carbon chains or carbon rings are still found in Organic Pigment molecules.	Metal cations are present in an array shape with nonmetallic anions in the molecules of Inorganic Pigments. This arrangement prevents the pigments from dissolving in the solvent or the plastic.
4.	Organic pigments have a high index of refraction	Inorganic pigments have significantly lower values
5.	Organic pigments are transparent	Most inorganic pigments are opaque

Table 1.
Difference between Organic & Inorganic Pigments.

5.2.1 White pigments

They are of various types. The composition, properties and applications of some white pigments can be summarised as follows;

5.2.1.1 White lead {$2PbCO_3.Pb(OH)_2$}

The composition of the Lead Carbonate is 68.9% and Lead Hydroxide is 31.1%. It is used in the manufacture of paints.

Properties:

- Easily applied
- High covering power
- Toxic in nature
- Yellows badly on exposure to the atmosphere
- Soluble in alkali and paints

5.2.1.2 Sublimed white lead (basic sulphate)

The composition of Lead Sulphate is 75%, Lead Oxide 20% and zinc oxide 5%. It is used in the manufacture of paints.

Properties:

- High specific gravity and refractive index
- Slow chalking out of the film producing a rough surface

5.2.1.3 Zinc oxide (ZnO)

The composition of Zinc Oxide is 100%. It is opaque to UV light and thus protects us from UV, prevents chalking.

Properties:

- Brilliantly white having excellent texture
- Causes no discolouration even in contact with CO_2 gas
- More durable in combination with white lead

5.2.1.4 Lithopone

The composition of lithopone is 28.30% zinc oxide and 72.70% barium sulphate. It is widely used for the cold water paints, traffic plants, floor covering and oilcloth industry.

Properties:

- Extremely fine and cheap pigment
- Good hiding power

- Not as durable as white lead and zinc oxide

5.2.1.5 Titanium dioxide

It consist of $TiFeO_3$, TiO_2 (iliminite+rutile). It is widely used in paints, in paper and textiles.

Properties:

- High capacity and hiding power
- High oil-absorbing capacity
- Spreading power is almost double as that of white lead
- No tendency of chalking

5.2.2 Blue pigments

Ultramarine Blue and Cobalt Blues are the most widely used blue pigments. The composition, properties and applications of these can be summarised as follows:

5.2.2.1 Ultramarine blue

There are three varieties of ultramarine namely blue, white and green. It is used as bluing in laundering to neutralise the yellowish tone in cotton and linen fabrics. White Ultramarine Blue: $Na_5Al_3Si_3SO_{12}$, Green Ultramarine Blue: $Na_5Al_3Si_2S_2O_{12}$, Blue Ultramarine Blue: $Na_5Al_3Si_2S_3O_{12}$.

Properties:

- Silicate skeleton have a potential influence on the colour
- Colour is because S present is in the form of polysulphide.

5.2.2.2 Cobalt blues

The composition of cobalt blues is 30–35% Co_3O_4 and 65–70% Al_2O_3. It is used in the manufacture of blue paints, inks, carbon paper and carbon ribbons.

Properties:

- Very expensive and are not used in paints for ordinary purposes.

5.2.3 Red pigments

These are one of the oldest-pigments. They are mainly used for inhibiting rusting of iron and steel structures. The composition, properties and applications of these can be summarised as follows:

5.2.3.1 Red lead

It consists of Pb_3O_4 and PbO, used for a primary coat on structural steel and in imparting red colour to the glass for making bangles.

Properties:

- Bright-red powder with a high specific gravity
- Excellent covering power
- Inhibits corrosion

5.2.3.2 Synthetic iron (Fe_3O_4)

It is widely used in domestic paints, enamels, floors and paints.
Properties:

- Has a dark brilliant colour
- High covering power and tinting strength

5.2.4 Green pigments

There are two types of commonly used green pigments. The composition, properties and applications of these can be summarised as follows:

5.2.4.1 Chrome green (Cr_2O_3)

It is generally used as green pigments.
Properties:

- High power of oil absorption.
- It has disadvantages such as lack of brilliancy and opacity

5.2.5 Chromium oxide or guignet's green [$Cr_2O(OH)_4$]

It is used as paint for metal surface and as non-fading green for washable distempers.
Properties:

- Have high covering power
- High corrosion inhibition capacity

5.2.6 Black pigments

They have good tinting strength and as well as high hiding power. The composition, properties and applications of these can be summarised as follows:

5.2.6.1 Natural black oxide

The composition of Fe_2O_3 is 94–95%. It is used in making paints for priming metal.
Properties:

- Oil absorption power is 10–15 kg of linseed oil per 100 kg of Pigment

5.2.6.2 Precipitated black iron oxide

It is used in cement emulsions and water paints. It has a high hiding power.

5.2.6.3 Carbon black/furnace black

It is used in making waterproof paints.
Properties:

- Increases the life of paints
- Good tinting strength
- Not affected by light, acids and alkalies

5.2.6.4 Lamp black

It is used in making black pigments.
Properties:

- Good tinting strength
- Resistant to high temperature

5.2.7 Yellow pigments

The composition, properties and applications of these can be summarised as follows:

5.2.7.1 Ochre

It consists of naturally occurring yellow Fe_2O_3. It is used in the paint industry.
Properties:

- Fast to light
- Inert to chemical action

5.2.8 Chrome yellow

It is used in making yellow paints.
Properties:

- Great opacity
- High brilliance
- High hiding power
- High tinting strength

5.2.9 Toners

Insoluble organic dyes are known as tonners. These can be used as pigments, quite durable and have high colouring power. Various dyes such as para red, Hansa Yellow G (lemon yellow), Hansa Yellow 10G (primrose yellow) etc. have been used as toners in the pigment industry.

5.2.10 Metallic powders as pigments

The powdered form of some metals, as well as some alloys, are used as pigments. Finely powdered aluminium and bronze have been used as pigments in lacquers. Pigments containing finely powdered zinc have been used for protective coatings on iron and steel to protect them from atmospheric corrosion [17–30].

6. Conclusions

This comprehensive chapter on structure and properties of dyes and pigments including some specific applications has been discussed. Moreover, recently the consumers have become very much conscious about the environment, renaissance of eco-friendly products and process like dyeing textiles with natural dyes, which has thus become also important now. Thus, revival of natural dye application on textiles and summary of earlier researches on standardisation of its method of extraction, mordanting, dyeing process variables and even natural finishing, etc. have been elaborated in this chapter. Thus this part has become a unique comprehensive chapter for information on structure and properties of dyes and pigments.

Conflict of interest

There is no conflict of interest.

Author details

Ashok Kumar[1], Utkarsh Dixit[1], Kaman Singh[1*], Satya Prakash Gupta[2] and Mirza S. Jamal Beg[3,4]

1 Laboratory of Physical Chemistry, Department of Chemistry, Babasaheb Bhimrao Ambedkar University (A Central University), Lucknow, Uttar Pradesh, India

2 Prasad Polytechnic Lucknow, Kanpur Road, Banthara, Sarai Sahjadi, Lucknow, Uttar Pradesh, India

3 University Institute of Engineering Technology, Babasaheb Bhimrao Ambedkar University (A Central University), Lucknow, Uttar Pradesh, India

4 Integrated Basic Sciences, Babasaheb Bhimrao Ambedkar University (A Central University), Lucknow, Uttar Pradesh, India

*Address all correspondence to: drkamansignh@yahoo.com

References

[1] Zanzoni A, Montecchi-Palazzi L, Quondam MX. Mint: A molecular interaction database. FEBS Letters. 2002;513:135–140. DOI: 10.1016/s0014-5793(01)03293-8

[2] Zanzoni A, Montecchi-Palazzi L, Quondam MX. Mint: A molecular interaction database. FEBS Letters. DOI: 10.1016/s0014-5793(01)03293-8

[3] Luque A, Hegedus S. Handbook of Photovoltaic Science and Engineering. 2nd ed. Chichester: Wiley; 2011. 1132 p. DOI: 10.1002/9780470974704

[4] Tiwari SC, Bharat A. Natural dye-yielding plants and indigenous knowledge of dye preparation in Achanakmar-Amarkantak Biosphere Reserve, Central India. Natural Product Radiance. 2008;7(1):82–87

[5] Rahman Bhuiyan MA, Islam A, Ali A, Islam MN. Color and chemical constitution of natural dye henna (Lawsonia inermis L) and its application in the coloration of textiles. Journal of Cleaner Production. 2017; 167:14–22. DOI: 10.1016/j.jclepro.2017.08.142

[6] Vankar PS, Shanker R, Dixit S, Mahanta D, Tiwari SC. Sonicator dyeing of modified cotton, wool and silk with Mahonia napaulensis DC and identification of the colorant in Mahonia. Industrial Crops and Products. 2008;27:371–379

[7] Kamel MM, El-Shishtawy RM, Yussef BM, Mashaly H. Ultrasonic assisted dyeing III. Dyeing of wool with lac as a natural dye. Dyes and Pigments. 2005;65:103–110

[8] Bouzidi A, Baaka N, Salem N, Mhenni MF, Mighri Z. Limoniastrum monopetalum stems as a new source of natural colorant for dyeing wool fabrics. Fibers and Polymers. 2016;17(8):1256–1261. DOI: 10.1007/s12221-016-5664-z

[9] Komboonchoo S, Bechtold T. Natural dyeing of wool and hair with indigo carmine (C.I. Natural Blue 2), a renewable resource based blue dye. Journal of Cleaner Production. 2009;17: 1487–1493

[10] Naz S, Bhatti IA, Adeel S. Dyeing properties of cotton fabric using un-irradiated and gamma irradiated extracts of Eucalyptus camaldulensis bark powder. Indian Journal of Fibre and Textile Research. 2011;36(2):132–136

[11] Samanta AK, Agarwal P, Datta S. Dyeing of jute and cotton fabrics using jackfruit wood extract: Part I—Effects of mordanting and dyeing process variables on colour yield and colour fastness properties. Indian Journal of Fibre and Textile Research. 2007;32(4):171–180

[12] Garifield S, Mauve WW. Natural Dyes for Textiles. New York.: Norton and Company; 2001

[13] Samanta AK, Agarwal P. Application of mixture of red sandal wood and other natural dyes for dyeing of jute fabric-studies on dye compatibility. International Dyer. 2008;192(3):37–41

[14] Chattopadhyay SN, Pan NC, Roy AK, Saxena S, Khan A. Development of natural dyed jute fabric with improved colour yield and UV protection characteristics. The Journal of the Textile Institute. 2013;104(8): 808–818. DOI: 10.1080/00405000.2012.758352

[15] Daberao AM, Kolte PP, Turukmane RN. Cotton dying with natural dye. International Journal ofResearch and Scientific Innovation (IJRSI). 2016;III(VIII):157–161.ISSN: 2321–705

[16] Luque A, Hegedus S, editors. Handbook of Photovoltaic Science and Engineering. 2nd ed. Chichester: Wiley;

2011. 1132 p. DOI: 10.1002/978047974704

[17] Ceccaroli B, Lohne O. Solar grade silicon feedstock. In: Luque A, Hegedus S, editors. Handbook of Photovoltaic Science and Engineering. 2nd ed. Chichester: Wiley; 2011. p. 169–217. DOI: 10.1002/978047974704.ch5

[18] Kajihara A, Harakawa T. Model of photovoltaic cell circuits under partial shading. In: Proceedings of the IEEE International Conference on Industrial Technology (ICIT '05); 14–17 December 2005; Hong Kong. New York: IEEE; 2006. p. 866–870

[19] Solarex. SX-40 & SX-50 Photovoltaic Modules [Internet]. 1999. Available from: http://www.trichord-inc.com/pricing/frames/content/solar_power.pdf [Accessed: YYYY-MM-DD]

[20] DenHerder T. Design and simulation of PV super system using Simulink [thesis]. San Luis Obispo: California Polytechnic State University; 2006.

[21] Dr. S. N. Dhawan. Pradeep's Organic Chemistry Vol.2

[22] Optional Module-2, Chemistry and Industry: Wiley; 2011.p.103–118.DOI: 10.1002/978047974704.ch34

[23] Anon (1996). Ecological and toxicological association of dyes and pigments manufacturers, textile chemists and colorist, "german ban of use of certain azo compounds in some consumer goods: ETAD Information Notice No. 6", Vol. 28 (4), 11.

[24] Aspland JR (1997). Textile dyeing and coloration. Association of Textile Chemists and Colorists. pp. 3–310.

[25] Corbett J (2000) Hair Dyes. In Freeman HS, Peter AT, eds, Colorants for Non-textile Applications, Amsterdam: Elsevier Science, pp. 456–477.

[26] Longstaff E. An assessment and categorisation of the animal carcinogenicity data on selected dyestuffs and an extrapolation of those data to an evaluation of the relative carcinogenic risk to man. Dyes and Pigments. 1983;4:243–304.

[27] Maron DM, Ames BN. Revised methods for the Salmonella mutagenicity test. Mutat Res. 1983;113: 173–215.

[28] Prival MJ, Bell SJ, Mitchell VD. et al. Mutagenicity of benzidine and benzidine-congener dyes and selected monoazo dyes in a modified Salmonella assay. Mutat Res. 1984;136:33–47.

[29] Weisburger E (1978). Cancer-Causing Chemicals. In LaFond RE, ed, Cancer — The Outlaw Cell, American Chemical Society, pp. 73–86.

[30] Yusuf, M., Shabbir, M. & Mohammad, F. Natural colorants: historical, processing and sustainable prospects. Nat. Prod. Bioprospect. 7, 123–145 (2017). DOI:10.1007/s13659-017-0119-9

Section 3

Dye Waste Treatment

Chapter 9

A Brief Comparative Study on Removal of Toxic Dyes by Different Types of Clay

Ahmed Zaghloul, Ridouan Benhiti, Rachid Aziam, Abdeljalil Ait Ichou, Mhamed Abali, Amina Soudani, Fouad Sinan, Mohamed Zerbet and Mohamed Chiban

Abstract

Increasing amount of organic dyes in the ecosystem particularly in wastewater has propelled the search for more efficient low-cost bio adsorbents. Different techniques have been used for the treatment of wastewater containing toxic dyes such as: biological degradation, oxidation, adsorption, reverse osmosis, and membrane filtration. Among all these processes mentioned, adsorption with low cost adsorbents has been recognized as one of the cost effective and efficient techniques for treatment of industrial wastewater from organic and inorganic pollutants. Clays as material adsorbents for the removal of various toxic dyes from aqueous solutions as potential alternatives to activated carbons has recently received widespread attention because of the environmental-friendly nature of clay materials. This chapter presents a comprehensive account of the techniques used for the removal of industrial cationic and anionic dyes from water during the last 10 years with special reference to the adsorption by using low cost materials in decontamination processes. Effects of different adsorption parameters on the performance of clays as adsorbents have been also discussed. Various challenges encountered in using clay materials are highlighted and a number of future prospects for the adsorbents are proposed.

Keywords: adsorption, toxic dyes, Clays, wastewater treatment

1. Introduction

The treatment of industrial wastewater particularly loaded with dangerous dyes is considered among the global environmental issues and the concerns of researchers [1]. According to recent reports, over one million dyes are available commercially with an annual output of over $7x10^5$ tons [2]. The textile industry around the world consumes approximately 10^4 tons of dyes annually and discharges about 100 tons of dyes into wastewater every year [3]. These dyes are highly toxic, carcinogenic, and cause dire consequences for human health and the marine system. The removal of these toxic dyes from polluted water and wastewater is highly desirable in order to meet regulatory obligations for wastewater recycling or discharging into natural environments [4]. Currently, there is several physico-chemical and biological technologies in the use for the treatment of these polluted effluents namely, ion exchange, membrane separation, biological treatment and adsorption [5–10]. Physico-chemical processes such as ion exchange, electro dialysis and reverse osmosis are expensive, difficult to operate and require significant technologies. While in recent years, adsorption has continued to attract the attention of the researchers worldwide [11, 12] and appears to be an alternative, which has some advantages such as simple design, and ease of operation. Whereas, the biological treatment which is based on the microbial digestive metabolism, has the major drawback of the risk of microbiological contamination and a significant production of sludge, which poses problems of storage and handling [13, 14]. In the present work a comparative study between the capacities of raw, synthetic and modified clays for the removal of toxic dyes from aqueous solution has been given, with particular review of the main factors influencing the adsorption of dyes by clays such as pH of the solution, temperature and initial dye concentration on the adsorption capacities of the these clays.

2. Removal methods for toxic dyes

Currently, a number of different technologies and methods such as membrane separation, ion exchange, adsorption and biological methods are widely used for the removal of toxic anionic and cationic dyes from polluted water and wastewaters.

2.1 Membrane filtration

Membrane separation is a pressure driven process. Pressure-focused processes are generally divided into four overlapping classes of increased selectivity: microfiltration (MF), ultrafiltration (UF), nanofiltration (NF) and hyperfiltration or reverse osmosis (RO). Microfiltration can be used to remove bacteria and suspended solids with pore sizes from 0.1 to microns. Whereas, Ultrafiltration eliminates colloids, viruses, and some proteins by pores from 0.0003 to 0.1 microns. Nanofiltration is based on physical rejection based on molecular size and charge. The pore sizes are between 0.001 and 0.003 microns [15]. Reverse osmosis has a pore size approximately 0.0005 microns and can be used for desalination. High pressures are needed to makepass water through the membrane from a concentrated solution to dilute. Shih [16] has studied the elimination of dyes on membrane and explored the parameters that could influence the efficiency of toxic dyes removal by membrane technologies such as parameters source, membrane type and membrane process.

2.2 Ion exchange

Ion exchange has been widely used to remove dyes due to its many advantages, such as high processing capacity, speed and increasing the efficiency of dye retention [17]. Ion exchange resin, either natural or resin solid synthetic, has the specific ability to exchange its cations with dyes in wastewater. Among materials most used in the ion exchange process, synthetic resins: are commonly preferred because they are effective in virtually removing dyes in solution [18].

2.3 Adsorption

Adsorption is a process in which solids are used for removing organic and inorganic substances from either gaseous or liquid solutions. The phenomena

Removal process		Advantages	Disadvantages
Chemical methods			
	Photo-catalyst	Low cost operational and economically feasible	Some photo catalyst degrades into toxic by-products.
	Ozonation	No sludge generation	Operational cost is very high, half life is short (20 min)
Biological methods			
	Anaerobic degradation	By-products can be used as energy	Resources under aerobic conditions require more treatment and yield of methane and hydrogen sulphide
	Aerobic degradation	Operational cost is low and effective in removal of azo dyes	Provide suitable environment for growth of microorganisms and very slow process
Physico-chemical methods			
	Adsorption	High adsorption capacity for all dyes.	Low surface area and high cost of some adsorbents.
	Ion exchange	No loss of sorbents	For disperse dyes not effective
	Membrane filtration	Effective for dyes with high quality effluents	Production of sludge and suitable for treating low volume

Table 1.
Separation techniques and their advantages and disadvantages [21, 22].

driven adsorption are operative in most natural physical, biological, and chemical systems. The solid adsorbents widely used in the industries for the removal of these pollutants from industrial wastewater are diverse such as activated carbon, metal hydrides and synthetic resins.

The adsorption process involves the separation of a substance from one phase by retaining it on the surface of another. The physical adsorption is mainly due to weak interactions such as van der Waals bonds and the electrostatic forces created between the adsorbate and the atoms which make up the surface of the adsorbent. The capacity of this process depends some parameters namely, adsorbent properties, adsorbate chemical properties, temperature, and pH of the medium. It should be noted that even if the adsorbents are available, they are still expensive and few of them are selective. Therefore, over the last decades the research has been redirected towards the search for other improved materials which will meet certain requirements such as regenerative capacity, easy availability, and cost effectiveness. Consequently, clays adsorbents have drawn attention to many researchers and characteristics as well as application of many such adsorbents are reported [19, 20]. However, clays adsorbents are discussed herein after. A summary of advantages and disadvantages of some separation methods are presented in **Table 1**.

3. Clay materials used as adsorbents for dye treatment

3.1 Layered double hydroxides (LDHs)

Layered double hydroxides (LDHs) are intensively studied because of their high anionic exchange capacity [23], reuse, larger surface area, porosity, and fundamental properties [24]. They have advantages over commercially available adsorbents in terms of low cost, high adsorption properties, and non-toxicity. The use of LDH

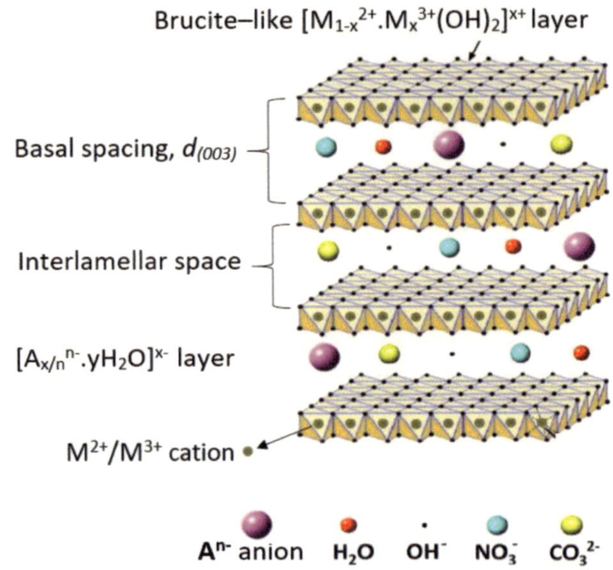

Figure 1.
A schematic representation of the LDH structure [28, 29].

could bring significant economic and environmental benefits to the wastewater treatment industries. Previous works [25] have proven their usefulness as adsorbents for the removal of some organic and inorganic pollutants from polluted water and wastewater. Among the different types of LDH material, a species similar to hydrotalcite, a compound consisting of a double compound of MgAl-LDH hydroxides with carbonate as interlayer anions, is commonly used for various applications [26, 27]. A schematic representation of the general structure of the LDH structure is given in **Figure 1**.

The effectiveness of these compounds in the treatment of polluted water and in particular organic pollutants such as textile dyes has already been demonstrated by various research teams [30–33]. Studies that have been done by Elmoubarki *et al.* [34] demonstrate the effectiveness of Layered double hydroxides based on Mg and Fe for the removal of methyl orange. Shan et al. [35] has studied the trapping of Congo red on carbonated HDL Mg (II) /Al (III), this study showed that the materials are more efficient, and have a very high yield of Congo red elimination. **Table 2** shows some adsorbents of the lamellaires types, used for the removal of some toxic dyes from aqueous solutions.

3.2 Kaolinite

The kaolinite group is comprised of trioctahedral minerals such as, chrysotile, antigorite, cronstedite, and chamosite, dioctahedral minerals such as kaolinite, halloysite, dickite and nacrite. It is white and soft clay, composed primarily of the mineral kaolinite, a hydrated aluminum silicate. Commonly, the kaolinite structure group is known to be composed of silicate sheets (Si_2O_5) linked to aluminum oxide/hydroxide layers ($Al_2(OH)_4$) called gibbsitelayers [51]. Additionally, the primary structural unit of this group is a layer composed of one octahedral sheet condensed with one tetrahedral sheet. About the dioctahedral minerals, the octahedral sites are occupied by aluminum, while those of trioctahedral minerals are occupied by magnesium and iron. Kaolinite and halloysite are single layer structures. Furthermore, kaolinite, nacrite and dickite occur as plates; halloysite, which can have a single

Adsorbent	Dye	Reference
NiFe-LDH	Methyl orange	[36]
ZnMgAl–CO_3	Methyl orange	[37]
MgNiAl	Methyl orange	[38]
MgAl- LDH	Methyl orange	[39]
rGO/Ni/MMO	Methyl orange	[40]
ZnO/CuO/ɣ-Al_2O_3	Methyl orange	[41]
Fe_3O_4/ZnCr-LDH	Methyl orange	[42]
NiAl-LDH	Congo red	[43]
Mg–Al–LDH	Congo red	[44]
CaAl-NO_3	Congo red	[45]
Ni/Fe-CO_3	Congo red	[46]
Mg/FeCO_3	Congo red	[47]
Mg - Al - Cl	Congo red	[48]
Zn-Fe–LDH	Methylene blue	[49]
Mn-Fe–LDH	Methylene blue	[49]
Mg-Fe–LDH	Methylene blue	[49]
GOaerogels/MgAl	Methylene blue	[49]
Zn- Fe -CO_3	Indigo Carmine	[50]
Zn-Cr -CO_3	Indigo Carmine	[50]
Zn-Mn -CO_3	Indigo Carmine	[50]
Zn-Al -CO_3	Indigo Carmine	[50]

Table 2.
Adsorbents used for the removal of dyes.

layer of water between its sheets, occurs in a tubular form. It consists of feldspar and muscovite formed by the alteration of [51, 52], and is a layered silicate mineral composed of a tetrahedral sheet, bonded through oxygen atoms to an octahedral sheet of alumina, which are layered silicate minerals composed of one tetrahedral sheet, linked through the oxygen atoms to one octahedral sheet of alumina octahedra. Nacrite, kaolinite and dickite exist as plates, halloysite occurs in a tubular form, have a single layer of water between its sheets. Rocks having large amount of kaolinite are referred to as kaolin or china clay [53]. Kaolinite contains heterogeneous surface charge is a well-known fact. It is believed that its basal surface has a constant structural charge which is attributed to isomorphs substitutions of Si^{4+} by Al^{3+}. The charge on the edges is due to protonation or deprotonation of surface hydroxyl groups and so it depends on pH of solution. Adsorption can occur on flat exposed planes of silica and alumina sheets. It is least reactive clay. Kaolin has no side effects, no health problems till the fine dust particle is controlled, thus it is safe environmentally [54, 55].

3.3 Bentonite

The most common group of clay used in water treatment is bentonite. It is a low-cost, effective and eco-friendly adsorbent, and it is commonly impure clay consisting mostly of montmorillonite, although some may consist of the rare clay minerals such as, nontronite saponite, and beidellite. Montmorillonite structure is a layer of gypsum site sandwiched between two sheets of silica to form the structural unit [56, 57]. The substitutes are found mainly in the octahedral layer (Mg^{2+}, Fe^{2+}) and to a lesser extent in the silicate layer. The clay mineral group is mainly composed of a hydroxyl-aluminosilicate framework. As well as, the crystal structures of the clay minerals are composed of a combination of silica tetrahedral sheets and alumino octahedral. Apart of the trivalent Al^{3+} is substituted by Mg^{2+} or Fe^{2+} ions in some cases. In such cases, substitution is accompanied by the addition of alkaline metals like Na^{+} and K^{+} or alkaline earth metals like Mg^{2+} and Ca^{2+} to provide charge balance [57]. Stockmeyer et al. (1991) [58] have investigated the adsorption of some organic compounds from aqueous solutions by using organophilic bentonites. Phenol, diethyl ketone, nitroethane, aniline ethoxy acetic acid, maleic acid and hexadecyl pyridinium bromide were investigated as test organic compounds. The used organophilic bentonites vary in the degree of their total cation exchange capacity exchanged by organic counter ions [58].

4. Parameters influencing the adsorption of dyes by clays

4.1 Effect of pH

The pH value of the solution is an important factor for the adsorption, which influences the structure of the adsorbates and the surface properties of adsorbents [59]. Commonly, at low pH of solution, the adsorption capacity and percentage removal of anionic dyes from aqueous solution increases due to electrostatic forces between the anionic dye molecule and positive surface charge of adsorbent. There is an electrostatic attraction between the positively charged dye molecule and negatively charged adsorbent. Besides, at high pH, the removal efficiency of anionic dyes decreases with increase in pH [60–62]. Whereas, the percentage removal and the amount adsorbed of cationic dyes at high pH increases because positive charges on the dye molecules ensured that they are attracted by anionic adsorbent, so there are electrostatic attractions between the negative surface of adsorbent and positive charges of dye molecules [63–66].

The previously reported literature indicates that optimized pH depends upon nature of dye and type of clay. Zaghloul et al. [67] studies the effect of pH on the removal of methyl orange (anionic dye) from aqueous solutions using synthetic clay type MgAl-LDH. It was observed in the range of pH from 2 to pH = 10, the percentage removal is very important (98%), due to the electrostatic forces between the anionic dye and the positively charged H^{+} surface of the synthetic clay as an adsorbent. At higher pH adsorption capacity and percentage removal of dyes decreased and this decrease has been explained by the fact that at pH = 10, number of hydroxyl ions is more and hence the competition between OH^{-} ions and anionic molecules for active adsorption sites. A similar investigation of the textile dye removal by another adsorbents has been reported by other researchers in the literature [59, 68].

4.2 Effect of temperature

Temperature is a crucial parameter in adsorption reactions. In general, the influence of temperature on the adsorption kinetics is very variable. Adsorption may increase, decrease, or remain constant with increasing temperature. Some studies have shown that a decrease in toxic dyes retention by clay materials is often accentuated by an increase in temperature [69]. Other works have shown that the adsorption of industrial dyes on different adsorbents increases with increasing temperature [70, 71]. Elmoubarki et al. [72] studied the adsorption of methylene blue and methyl orange on synthetic clays of Ni-LDH, Mg-LDH types, in a temperature range (30 to 50 °C). They have observed that the quantities of MB and MO adsorbed as a function of temperature increase with the increase in temperature. On the other hand, the research carried out by Zaghloul et al. [67] on the adsorption of methyl orange by MgAl-LDH (2:1) have shown that the temperature increase from 30 to 35 °C disadvantages the adsorption of methyl orange onto LDH.

4.3 Effect of initial dye concentration

The initial concentration of adsorbate and adsorbent has a great importance in batch and fixed bed column adsorption experiments, because it depends on the nature of the system used. Sureshkumar et al. [73] showed that an increase in the retention of dyes by clays is promoted by an increase in the initial concentration of these dyes. Likewise, the others work [74] shows that the retained concentration of dyes (methyl orange, crystal violet, blue acid) increases with the initial concentration of these dyes. In our previous reports [67], we have studied the effect of initial concentration of methyl orange by MgAl-LDH (2:1), it was observed that the amount of methyl orange adsorbed as a function of temperature increases with increase in concentration. The same remark was recorded by Krika and Benlahbib [75] during the retention of methyl orange by cork powder. They have demonstrated that adsorption process is an effective method because of its efficiency, capacity, and applicability on large scale dye-removal, as well as the potential for regeneration, recovery, and recycling of adsorbents.

5. Equilibrium studies

Equilibrium studies explore the relationship between adsorbent and adsorbate which is described by adsorption isotherms [76]. The adsorption isotherm studies are important both a theoretical and a practical point of view. Further, isotherm data must precisely fit different isotherm models to find an appropriate model that can be used for the design process [77, 78]. The obtained parameters from the different models provide important information on the adsorption mechanisms, the surface properties and affinities of the adsorbent. Several models have been published in the literature to describe experimental equilibrium data of adsorption isotherms. The most famous adsorption models for single-solute systems are Freundlich, Langmuir, Redlich–Peterson, Radke–Prausnitz, Koble–Corrigan, Temkin, Dubinin–Radushkevich (D–R), BET (Brunauner, Emmett, Teller), Sips and Generalized isotherms.

Langmuir adsorption isotherm assumes that the solid surface has a finite number of identical sites which shows homogeneous surfaces. Langmuir equation may be represented as [79]:

$$q_e = q_L \frac{K_L C_e}{1 + K_L C_e} \tag{1}$$

where q_e (mg/g) is the amount of the dye adsorbed per unit weight of clay at equilibrium, C_e (g/L) is the equilibrium concentration of dye in the solution, q_L (mg/g) is Langmuir maximum adsorption capacity and K_L (L/g) is Langmuir constant related to a free energy of adsorption.

The Freundlich isotherm is an empirical equation that assumes that the adsorption surface becomes heterogeneous during the adsorption process. The Freundlich isotherm is expressed by the following Equation [80].

$$q_e = K_F C_e^{1/n} \quad (2)$$

where, q_e (mg/g) is the amount of the dye adsorbed per unit weight of clay; C_e (g/L) is the equilibrium concentration of the dye in the bulk solution; K_f is Freundlich constant, which is a comparative measure of the adsorption capacity for the clay, and n_f is an empirical constant related to the heterogeneity of the material surface.

Zaghloul et al. [67] use a synthetic clay type MgAl-HDL for the removal of methyl orange. The Langmuir and Freundlich models were applied to the experimental data. The results indicated that the Langmuir isotherm fully describes the nature and sorption mechanism of methyl orange (MO) on the synthetic clay used. The adsorption capacity of MgAl-LDH calculated from the Langmuir model was found to be 1250 mg. g^{-1}. The value of 1/n at 298 K was 0.34 (i.e. $(1 < n < 10)$ indicating that the adsorption system solid–liquid studied was favorable. Bentahar et al. [81] have used clay minerals like bentonite, kaolin and zeolite for removal of Congo red dye from aqueous solutions. The results of both Freundlich and Langmuir models indicates that zeolite and bentonite were best described by the Freundlich model, however Langmuir model provided a better fit on the experimental data of kaolin with high R^2 value ($R^2 = 0.98$).

6. Kinetic studies

Kinetic studies are very important in regards of adsorption studies because they can describe the adsorption rate and provide valuable data for understanding the mechanism of adsorption reactions [82]. In order to understand the behavior of the adsorbent and to investigate the controlling mechanism of the adsorption procedure, the pseudo first-order, pseudo second order and intraparticle diffusion models are useful to check the kinetic information [83]. To explore the mechanism of adsorption regarding the adsorptive removal of methylene bleu (MB) by using amino-functionalized attapulgite clay nanoparticle Zhou et al. [84] fitted the experimental data to pseudo-first-order and pseudo-second-order kinetic models. The correlation coefficients of the pseudo second-order kinetic model were relatively greater than those of the pseudo first order kinetic model, implying that the MB adsorption can be described more appropriately by the pseudo-second-order model. Therefore, it can be concluded that a large number of vacant surface sites were available for adsorption during initial stage.

The reviewed research articles regarding kinetic studies show that pseudo-second order kinetic model is more suited to the experimental data compared to other models, however; depending upon the reaction other kinetic models also show correlation to the data.

7. Thermodynamic studies

Thermodynamic investigations are another important parameter of adsorption studies. The determination of thermodynamic parameters is an essential means of describing the energetic mechanism which operates in an adsorbent / adsorbate system during the adsorption process. For thermodynamic studies, the adsorption experiment should carry out at different temperature conditions and calculated parameters included standard enthalpy (ΔH°), standard entropy (ΔS°), and standard Gibbs free energy (ΔG°) [59].

$$K_d = \frac{q_e}{C_e} \tag{3}$$

$$\Delta G^\circ = -RTLnK_d \tag{4}$$

$$LnK_d = \frac{\Delta S^\circ}{R} - \frac{\Delta H^\circ}{RT} \tag{5}$$

Fan et al. [85] have studied the effect of temperature on the removal of methylene bleu by adsorption onto Mt-SB12. The thermodynamic study provides good information about the energetic changes related to adsorption process. The standard Gibbs free energy (ΔG°) values at different temperatures were found negative and standard enthalpy (ΔH°) were positive. These results indicates a spontaneous and endothermic nature of the adsorption process. Furthermore, the positive values of ΔS° reflected an increase in randomness at the solid/solution interface during the adsorption of methylene bleu onto Mt-SB12 [86]. In the same context, Zaghloul et al. [67] studied the effect of temperature on the removal of methyl orange using synthetic clay (MgAl-LDH) at different temperatures 25, 30 and 35 °C. ΔG° values at all temperatures were found to be negative implied that the adsorption of MO on MgAl-LDH was thermodynamically feasible and spontaneous [87]. The negative values of ΔH indicated that the adsorption process was exothermic in nature, and the negative values of ΔS designated a greater order of reaction during adsorption of MO dye by MgAl-LDH, which may be attributed to the adherence of dye molecule with MgAl-LDH adsorbent resulting a decrease in the degree of freedom of the system solid/ liquid [72]. Summary of the thermodynamic studies shows that the sorption process may be exothermic or endothermic for dyes adsorption onto clays.

8. Comparison of adsorption capacities

In order to justify the validity of low cost adsorbent for removal of toxic dyes, its adsorption potential must be compared with other materials used for this purpose. The values of maximum adsorption capacities in term of percentage removal of textile dyes onto different adsorbents reported in the literature are given in **Table 3**. The direct comparison of adsorption capacities of the adsorbents reported in the literature is difficult due to the varying experimental conditions employed in those studies. The adsorption capacity differences of toxic dyes uptake are ascribed to the properties of each adsorbent such as adsorbent structure, functional groups and surface areas.

9. Conclusion

For many decades, the raw, synthetic and modified clays have been considered low-cost and effective adsorbents, which have been successfully used for the adsorption of cationic and anionic dyes from polluted water and wastewater in the laboratory scale, although these several experiments but up to day few of researchers have focused on the use of these clays as adsorbent for the removal of industrial dyes from real effluents. The performance of different types of clays, whether raw,

Adsorbent	Adsorbate	% removal	Reference
NiFe-CO_3	Methyl orange	88	[59]
Mg/Fe-CO_3	Methyl orange	33	[88]
Montmorillonite	Methyl orange	90	[89]
Bentonite modified	Methyl orange	98	[90]
Ni/Al-CO_3	Congo Red	92	[91]
Mg/Al-CO_3	Congo Red	90	[92]
Natural kaolinitic clay	Congo Red	84	[93]
Ca-bentonite	Congo Red	95	[94]
Ghassoul	Methylene Blue	90	[95]
Algerian bentonite	Methylene Blue	91	[96]
Alginate/PVA–kaolin	Methylene Blue	99	[97]
Zeolite	Methylene Blue	88	[98]
Mg/Fe-CO_3	Malachite Green	86	[90]
Raw Moroccan clay	Malachite Green	—	[89]
Bentonite	Malachite Green	90	[98]
Moroccan clay	Basic Red 46	95	[98]
Raw clay (smectite)	Reactive Red 120	88	[99]
Raw clay (kaolinite)	Reactive Red 120	94	[99]

Table 3.
Adsorption capacities of some clay adsorbents for the removal of toxic dyes from water and wastewater.

synthetic or modified, was compared with regard to removing dyes based on some experimental parameters including pH, temperature and initial dye concentration. It was found that synthetic and modified clays provide a greater efficiency relating the removal of these organic pollutants.

Author details

Ahmed Zaghloul[1], Ridouan Benhiti[1], Rachid Aziam[1], Abdeljalil Ait Ichou[1], Mhamed Abali[1], Amina Soudani[2], Fouad Sinan[1], Mohamed Zerbet[1] and Mohamed Chiban[1*]

1 Department of Chemistry, Faculty of Science, Ibn Zohr University, Agadir, Morocco

2 Faculty of Applied Sciences, University Campus, Ait Melloul, Morocco

*Address all correspondence to: mmchiban@gmail.com

References

[1] Yao W., Yu S., Wang J., Zou Y., Lu S., Ai Y., Alharbi N.S., Alsaedi A., Hayat T., Wang X., Chem. Eng. J. 307 (2017) p. 476-486

[2] Robinson T., McMullan G., Marchant R., Nigam P., Bioresour. Technol. 77 (2001) p.247-255

[3] Yagub M.T., Sen T.K., Afroze S., Ang H.M., Adv. Colloid Interf. Sci. 209 (2014) p.172-184

[4] Han R, Ding D, Xu Y, Zo W, Wang Y, Li Y, Bioresour Technol. 99(2008) p.2938-46

[5] Chiban M., Soudani A., Zerbet M., Sinan F., Wastewater treatment processes. Chap. 10, Handbook of Wastewater treatment: Biological Methods, Technology and Environmental Impact, Editors: Cesaro J. Valdez and Enrique M. Maradona, Nova science publishers, Inc. New York, 2013, pp. 249-262, ISBN: 978-1-62257-591-6.

[6] Rijn J. V., Tal Y., Schreier H., J. Engg. 34 (2006) p. 364-376.

[7] Bhatnagar A., Sillanpaa M.,. Chem. Eng. J., 168 (2) (2011) p. 493-504

[8] Rozada F., Calv L.F., García A.I., Martín-Villacorta J., Otero M., Bioresour. Technol. 87 (2003) p. 221-230.

[9] Chiou M.S., Ho P.Y., Li H.Y., Dyes Pigments.60 (2004) p. 69-84.

[10] Janos P., Buchtova H., Rznarova M., Water Res. 37 (2003) p. 4938-4944.

[11] Wu X., Wang W., Li F., Khaimanov S., Tsidaeva N., Appl. Surf. Sci.389 (2016) p. 1003-1011.

[12] Wang S.L., Hseu R.J., Chang R.R., Chiang P.N., Chen J.H., Tzou Y.M., Colloids Surf. A: Physicochem. Eng. Aspects 277 (2006) p. 8-14

[13] Gou Y., Yang S., Fu W., Qi J., Li R., Wang Z., Xu H., Dyes Pigments. 56 (2003) p. 219-229.

[14] Ozdemir O., Armagan B., Turan M., Çelik M.S., Dyes Pigments 62 (2004) p. 49-60.

[15] Chiban M., Zerbet M., Carja G. Sinan F., Environ. Chem. Ecotox. 4(2012) p. 91-102

[16] Shih MC. Desalination. 172 (2005) p. 85-97.

[17] Kang S.Y, LEE J.U, Moon S.H, Kim K.W. Chemosphere. 56 (2004) p. 141-147

[18] Alyuz, B, Veli, S.. J. Hazard Mater. 167 (2009) p. 482-488

[19] Huang CP, Fu PLK.. J. Water Pollut. Control Fed 56 (1984) p. 233-242.

[20] Gimbel R, Hobby R, Discharge of arsenic and heavy metals from activated carbon filters during drinking water treatment. Wasser Rohrbau, 51 (2000) p. 15-16.

[21] Salleh M.A.M., Mahmoud D.K., Karim W.A.W.A., Idris A., Desalination 280 (2011) p. 1-13.

[22] Dawood S., Sen T.K., J. Chem. Process. Eng. 1 (2013) p. 1-7

[23] Ming Ni Z., Jie Xia S., Geng Wang L., Fang Xing F., Xiang Pan G., J. Colloid Interface Sci. 316 (2007) p.284-291.

[24] Li F., Wang Y., Yang Q., Evans D.G., Forano C., Duan X., J. Hazard. Mater. 125 (2005) p.89-95.

[25] Ait Ichou A., Abali M., Chiban M., Carja G., Zerbet M., Eddaoudi E., Sinan F., J. Mater. Environ. Sci. 5 (2014) p. 2444-2448.

[26] Aschenbrenner O., Guire P., Alsamaq S., Wang J., Supasitmongkol S., Al-Duri B., Styring P., J. Wood, Chem. Eng. Res. Des. 89 (2011) p.1711-1721.

[27] Reichle W.T., Solid State Ion. 22 (1986) p.135-141.

[28] Ait Ichou A., Élaboration et caractérisation d'argiles de type hydroxyde double lamellaire : Application à l'adsorption des ions métalliques en milieu aqueux, thesis of Ibn Zohr University, Morocco, 2019.

[29] Richetta M, Medaglia PG, Mattoccia A, Varone A, Pizzoferrato R (2017) Layered Double Hydroxides: Tailoring Interlamellar Nanospace for a Vast Field of Applications. J Material Sci Eng 6: 360.

[30] Krikaa F., Benlahbiba O.F., Desalin. Water Treat. 53 (2015) p.3711-3723

[31] Mekatel E., Amokrane S., Aid A., Nibou D., Trari M., Comptes Rendus Chimie .18 (2015) p.336-344

[32] Zheng Y.M., Li N., De Zhang W., Colloids Surf. A. Physicochem. Eng. Asp, 415 (2012) p. 195-201

[33] Ni Z., Xia S., Wang L., Xing F., Pan G., J. Colloid Interface Sci., 316 (2007) p. 284-291

[34] Kang H. T., Lv K., Yuan S. L., Appl. Clay Sci, 72 (2013) p.184-190.

[35] Ran-ran Shan , Liang-guo Yan, , Yan-ming Yang, Kun Yang , Shu-jun Yu , Hai-qin Yu , Bao-cun Zhu , Bin Du, J. Ind. Eng. Chem. 21 (2015)p. 561-568

[36] Zubair M., Jarrah N., Ihsanullah, A. Khalid, M. Manzar, T. Kazeem, M. Al-Harthi A., J. Mol. Liq. 249 (2018) p. 254-264

[37] Zheng Y.M., Li N., De Zhang W., Colloids Surf. A. Physicochem. Eng. Asp, 415 (2012) p. 195-201

[38] Zaghouane-Boudiaf H., Boutahala M., Arab L., Chem. Eng. J. 187(2012) p.142-149

[39] Aisawa S., Hirahara H., Uchiyama H., Takahashi S., Narita E., J. Solid State Chem., 167 (2002) p.152-159.

[40] Yang Z., Ji S., Gao W., Zhang C., Ren L., Weei W., Zhang Z., Pan J., Liu T., J. Colloid Interface Sci. 408 (2013) p. 25-32,

[41] Hassanzadeh-Tabrizi S.A., Motlagh M.M., Salahshour S., Appl. Surf. Sci. 384 (2016) p. 237-243.

[42] Chen C D., Li Y., Zhang J., Li W., Zhou J., Shao L., Qian G., , J. Hazard. Mater. 243 (2012) p. 152-160

[43] Bharali D., Deka R.C., Adsorptive removal of congo red from aqueous solution by sonochemically synthesized NiAl layered double hydroxide, J. Environ. Chem. Eng., , 5 (2017) p. 2056-2067.

[44] Lafi R., Charradi K., Djebbi M.A., Amara A.B.H., Hafiane A., Adv. Powder Technol. 27 (2016) 232-237.

[45] Nguyen Thi H. T., Nguyen D.C., Nguyen T.T., Tran V.T., Nguyen H.V., Bach L.G., Vo D.V.N., Nguyen D.H., Thuan D.V., Do S.T., Nguyen T. D., Key Engineering Materials, 814 (2019) p 463-468

[46] Ayawei N., godwin J. and wankasi D., Int. J. Chem. Sci. 13(3), (2015) p. 1197-1217

[47] Ayawei, N, Angaye S.S. Wankasi, D, Int. J. Appl. Sci. 7 (2017) 83-92.

[48] Selim Y., J. Disper. Sci. Technol. 7 (2017) 83-92.

[49] Shi Z., Wang Y., Sun S., Zhang C., Wan H., Water Sci. Technol. 81 (2020) p. 2522-2532

[50] Bouteraa S., Boukraa F. Saiah D., Hamouda S., Bettahar N., Bulletin of Chemical Reaction Engineering & Catalysis, 15 (2020) p.43-54

[51] Chaari I., Medhioub M., Jamoussi F., J. Appl. Sci. Environ. Sanitat. 6(2011) p.143-148.

[52] Moore D.M., Reynolds R.C., Oxford university press, Oxford, 1989.

[53] Murray H.H., Applied Clay Mineralogy: Occurrences, Processing and Applications of Kaolins, Bentonites, Palygorskitesepiolite, and Common Clays, Elsevier (2006)

[54] Liu P., Zhang L., Sep. Purif. Technol. 58 (2007) 32-39

[55] Nandi B.K., Goswami A., Purkait M.K., J. Hazard. Mater. 161 (2009) p.387-395.

[56] Tunega D, Haberhauer G, Gerzabek MH, Lischka. Langmuir 18 (2002) p.139-147

[57] Zakaria RM, Hassan I, El-Abd MZ, El-Tawil international water technology conference (IWTC), Hurghada, 13 (2009) p 403-416

[58] Stockmeyer M, Kruse K, Clay Miner 26 (1991) p.431-434

[59] Zubair M., Jarrah N., Ihsanullah A., Khalid M., Manzar T., Kazeem M., Al-Harthi A., J. Mol. Liq. 249 (2018) p.254-264.

[60] Nadeem R., Manzoor Q., Iqbal M., Nisar J., J. Ind. Eng. Chem. 35(2016) p.185-194.

[61] Naeem H., Bhatti H.N., Sadaf S., Iqbal M., Appl. Radiat. Isot. 123(2017) p.94-101.

[62] Shoukat S., Bhatti H.N., Iqbal M., Noreen S., Microporous Mesoporous Mater. 239 (2017) p.180-189.

[63] Tahir M.A., Bhatti H.N., Iqbal M., J. Environ. Chem. Eng. 4 (2016) p.2431-2439.

[64] Tahir N., Bhatti H.N., Iqbal M., Noreen S., Int. J. Biol. Macromol. 94(2016) p.210-220.

[65] Rashid A., Bhatti H.N., Iqbal M., Noreen S., Ecol. Eng. 91 (2016) p.459-471.

[66] Tahir M., Iqbal M., Abbas M., Tahir M., Nazir A., Iqbal D.N., Kanwal Q., Hassan F., Younas U., , Acta Ecol. Sin. 37 (2017) p.207-212.

[67] Zaghloul A., Ait Ichou A., Benhiti R., Abali M., Soudani A., Chiban M., Zerbet M., Sinan F., Mediterr.J. Chem. 9(2) (2019) p.155-163

[68] Huang R., Adsorption of methyl orange onto protonated cross-linked chitosan, Arab. J. Chem. 10 (2017) p.24-32

[69] Boumchita S., Lahrichi A., Benjelloun Y., Lairini S., Nenov V., Zerrouq F., J. Mater. Environ. Sci. 7 (1) (2016) p.73-84

[70] Janos P., Buchtova H., Rznarova M., Water Res 37 (2003), p.4938-4944.

[71] Gou Y., Yang S., Fu W., Qi J., Li R., Wang Z., Xu H., Dyes Pigments, 56 (2003), p.219-229.

[72] Elmoubarki R., Mahjoubi F., Tonsadi A., Abdennouri M., Sadiq M., Qourzal S., Zouhri A., Barka N.., J. Mater. Res. Technol. 3(2017) p.271-283.

[73] Sureshkumar M.V, Namasivayam C., Colloids Surf. A: Physico chem. Eng. Aspects, 317(2008): p.277-283

[74] Aziam R., Chiban M., Eddaoudi H., Soudani A., Zerbet M. and Sinan F., Eur. Phys. J. Special Topics 226 (2017) p.977-992.

[75] Krikaa F., Benlahbiba O.F., Desalination and Water Treatment 53 (2015) 3711-3723

[76] Toor M., Jin B., J. Am. Chem. Soc, 38 (1916), p.2221-2295

[77] Chiban M. (2011) Élaboration et Évaluation d'un Nouveau Procédé d'Épuration des Eaux : Application à des solutions modèles et d'eaux usées domestiques et industrielles de la région d'Agadir. Editions Universitaires Européennes, ISBN: 978-613-1-58836-5, 272p.

[78] Chiban M., Zerbet M., Sinan F. (2012) Low-cost materials for phosphate removal from aqueous solutions. Chap.1, Handbook of Phosphates: Sources, Properties and Applications, Nova Science Publishers, Inc. New York, pp.1-42, ISBN: 978-1-61942-123-3.

[79] Huang R., Arab. J. Chem, (2017) 10: p.24-32

[80] Freundlich H.M.F., Phys. Chem, (1906) 57: p.385-47029.

[81] Bentahar S., Dbik A., El Khomri M., El Messaoudi N., Bakiz B., Lacherai A.,. Res. Chem. Chem. Eng. Biotechnol. Food Ind. 17 (2016) p.295

[82] Shamsayei M., Yamini Y., Asiabi H., J. Colloid Interface Sci (2018) 529: p.255-264

[83] Elass K., Laachach A., Alaoui A., Azzi M., Appl. Clay Sci. 54 (2011) p.90-96

[84] Zhou Q., Gao Q., Luo W., Yan C., Ji Z., Duan P., Colloid. Surf. A Physicochem. Eng. Asp.470 (2015) p.248-257.

[85] Fan H., Zhou L., Jiang X., Huang Q., Lang W., Appl. Clay Sci. 95(2014) p.150-158.

[86] Dalhat N., Haladu S.A., Jarrah N., Zubair M., Essa M.H., Ali S.A., J. Hazard. Mater, 342 (2018) p.58-68.

[87] Shan R., Yan L., , Yang K., , Yu S., , Hao Y., , Yu H., , Du B,., Chem. Eng. J. 2014, 252, p.38-46.

[88] Chen D., Chen J., Luan X., Ji H., Xia Z.. Chemical Engineering Journal 171 (2011) p.1150-1158

[89] Tahir S.S., Naseem R., Chemosphere 63(2006): p.1842-1848

[90] Bharalia D., Deka R.C., Environmental Chemical Engineering 2017

[91] Lafi R., Charradi K., Djebbi M.A., Amara A.B.H, Hafiane A Adv. Powder Technol. (2016) 27: p.232-237

[92] Nwokem N.C., Nwokem C.O., Ayuba A.A., Usman Y.O., Odjobo B.O., Ocholi O.J., Batari M.L., Osunlaja A.A., Arch Appl Sci Res(2012) 4: p.939-946

[93] Lian L, Guo L, Guo C . J Hazard Mater (2009) 161: p.126-131

[94] Elass K, Laachach A, Alaoui A, Azzi M Appl Ecol Environ Res 8(2010) p.153-163

[95] Bellir K, Bencheikh-Lehocine M, Meniai A-H. Int Renew Energy Congr (2010): p.360-367

[96] Abd El-Latif MM., El-Kady MF., Ibrahim AM., Ossman ME., J Am Sci (2010) 6: p.280-292

[97] Zhu J., Wang Y., Liu J., Zhang Y., Ind. Eng. Chem. Res. 53 (2014) p.13711-13717.

[98] Bennani-Karim A, Mounir B, Hachkar M, Bakasse M, Yaacoubi A . Can J Environ Constr Civ Eng(2011) 2: p.5-13

[99] Errais E, Duplaya J, Elhabiri M, Khodjac M, Ocampod R, Baltenweck-Guyote R, Darragi F. Colloids Surf A. 403(2012) : p.69-78.

Chapter 10

Preparation of Functionalized Hydroxyapatite with Biopolymers as Efficient Adsorbents of Methylene Blue

Hassen Agougui, Youssef Guesmi and Mahjoub Jabli

Abstract

In this study, we reported the synthesis of hydroxyapatite modified with biopolymers as λ-carrageenan and sodium alginate, which could be used as effective adsorbents of cationic dyes. Evidence of chemical modification was proved through chemical analysis, Fourier Trans-form Infrared spectroscopy, powder X-ray diffraction, scanning electron microscopy, and specific surface area. The adsorption process was studied using methylene blue as representative cationic dye. The adsorbed quantity reached, at equilibrium, 142.85 mg/g and 98.23 mg/g using hydroxyapatite-sodium alginate and hydroxyapatite-(λ-carrageenan), respectively. However, it does not exceed 58.8 mg/g in the case of the unmodified hydroxyapatite. The adsorption of methylene blue using hybrid materials complied well with the pseudo-second-order suggesting a chemi-sorption. Freundlich and Langmuir isotherm described well the adsorption mechanism of the hydroxyapatite-(λ-carrageenan) and hydroxyapatite-sodium alginate, respectively. The high capacities of MB removal obtained in this study suggest the potential use of these materials in the treatment from wastewaters.

Keywords: hydroxyapatite, biopolymer alginate, dye, kinetic

1. Introduction

Contaminated waters can be successfully treated using inexpensive adsorbents. In this sense, many biopolymers were proposed including cellulose [1–3], chitosan [4, 5], chitin [6, 7], etc. Hybrid materials have attracted a particular attention. The interaction between calcium hydroxyapatite and biopolymers has been the subject of many studies such as carboxymethyl cellulose [8], polygalacturonic acid [9], collagen [10], Agar-Agar [11], polycaprolactone [12], banana peel [13], chitosan [14], and gelatin [15]. Currently, the application of hydroxyapatite modified by biopolymers for immobilization of various pollutants has been considered as a promising pollution control technology [16–20]. For example, Huijuan and colleagues [21] have, recently, reported interesting results about the preparation of hydroxyapatite-Chitosan composite and its efficiency for the removal of Congo red dye from aqueous solution.

The results indicate that the kinetic and isotherm studies showed that pseudo-second-order model and Langmuir model could well describe the adsorption behavior, while thermodynamic investigation of Congo red adsorption by hydroxyapatite-Chitosan composite confirmed a spontaneous adsorption. It has been demonstrated that this composite is an effective and low-cost adsorbent for the dye-polluted water purification [21].

In the same framework, the present work describes the synthesis of hybrid compounds CaHAp-Alginate and CaHAp-(Carrageenan) as an adsorbent using a facile method by varying the content of the bio-polymer. Evidence of interaction between hydroxyapatite and Alg or (λ-Carr) was confirmed using various techniques including FT-IR, SSA, DRX and SEM. The factors that influence the dye uptake by the prepared adsorbents were also investigated and discussed.

2. Experimental

2.1 Synthesis of hydroxyapatite (CaHAp1 and CaHAp2)

Hydroxyapatite was synthesized via co-precipitation method [22, 23]. As starting reagents, analytical grade $CaCl_2$, Na_2HPO_4, $Ca(NO_3)_2.4H_2O$, $(NH_4)_2HPO_4$ and NH_4OH were used. The chemicals equations that describes the reactions is given as follow:

CaHAp1

$$10Ca(NO_3)_2.4H_2O+6(NH_4)_2HPO_4+8(NH_4OH)\rightarrow Ca_{10}(PO_4)_6(OH)_2+20(NH_4^+)+20(NO_3^-)+46H_2O \quad (1)$$

CaHAp2

$$10CaCl_2+6Na_2HPO_4+2H_2O\rightarrow Ca_{10}(PO_4)_6(OH)_2+12(Na+,Cl-)+8(H+,Cl^-) \quad (2)$$

An aqueous solution of 250 mL of $Ca(NO_3)_2.4H_2O$ or $CaCl_2$ (0.2 M) was added drop-wise to 150 mL of $(NH_4)_2HPO_4$ or Na_2HPO_4 (0.2 M) solution under N_2 bubbling with a stoichiometric ratio of Ca/P = 1.67.

The pH was adjusted to 10 by adding NH_4OH solution gradually. After the complete addition, the suspension was matured for 72 h at room temperature under magnetic stirring, the resultant precipitate was filtered, washed with distilled water, and dried overnight at 80°C.

2.2 Synthesis of hydroxyapatite modified by sodium alginate CaHAp1-Alg

A similar procedure of CaHAp1 was adopted to synthesize the modified hydroxyapatite by sodium alginate CaHAp1-(Alg), except for the fact that the $(NH_4)_2HPO_4$was prepared with x mass of sodium alginate (5%, 10% or 20% of the total mass of $Ca(NO_3)_2.4H_2O$ in the starting solution).

2.3 Synthesis of hydroxyapatite modified by sodium alginate CaHAp1-(λ-Carr)

CaHAp2-(Carr) was synthesized, following a similar procedure of CaHAp2, except for the calcium containing solutions that were prepared by mixing 0.05 moles of $CaCl_2$ with x mol (x = 0.0025, 0.005 and 0.01) of (λ Carr).

Figure 1.
Structure of methylene blue.

In the text, Hydroxyapatite modified with (λ-Carr) is abbreviated as (CaHAp-(Carr) n), where n represents the mass ratio $\frac{m(Carr)}{m(calcium)} \times 100.$

2.4 Adsorption experiments

MB (its structure is given in **Figure 1**) solution was prepared by dissolving the calculated powder dye in distilled water with different required concentrations. NaOH and acetic acid solutions were used for adjusting the p*H* of the dye solution. For the adsorption experiments, MB solution was mixed with modified hydroxyapatite. MB concentrations were measured before and after experiments using a double beam UV–vis spectrophotometer (UV-1601 Shimadzu, Japan) at 664 nm. The adsorbed amount of dye onto modified hydroxyapatite (q mg/g) was calculated using the following equation [24]:

$$q_e = (C_O - C_e) x \frac{V}{m} \quad (3)$$

Where C_o and C_e are the initial and equilibrium dye concentration (mg/L), V is the solution volume (mL) and m is the weight of used modified hydroxyapatite sample (g) for the adsorption.

3. Characterization techniques

The FT-IR spectra were recorded on a Perkin Elmer model 597 using KBr pellet method in the 4000–400 cm^{-1} region. X-ray powder diffractograms were obtained at room temperature on a PANalytical X'Pert PRO MPD equipped with copper anticathode tube. The morphological observation of the synthesized samples was undertaken using a JEOL JSM-5400 scanning electron microscope. Specific surface area (SSA) measurements were performed by BET-method (adsorptive gas N_2, carrier gas He, heating temperature 100°C) using an Quantachrome Instruments, model: ASIM. LP2. The pH of the zero point charge (pH_{ZPC}) was determined by putting 0.15 g of adsorbent in a closed Erlenmeyer flask containing 50 mL of NaCl solutions (0.1 M). The initial pH of these solutions was adjusted by either adding NaOH (0.1 M) or HCl (0.1 M) and were then agitated for 48 h at 150 rpm at room temperature to reach equilibrium. The final pH of supernatant was, further, measured and the ΔpH = pH (final) – pH (initial) was plotted against the initial pH. The pH at which ΔpH was zero was taken as a pH_{ZPC}.

4. Results and discussion

4.1 Characterization of adsorbents

4.1.1 X-ray diffraction

X-ray diffractograms of hydroxyapatite-Alg and hydroxyapatite-(λ-Carr) are presented in **Figures 2** and **3**. For all compounds, the hydroxyapatite phase is conserved according to 00-024-0033 reference from the ICDD-PDF2 2003 database. However, in the case of CaHAp1-(Alg)20% new peaks appear at $2\theta = 26.60°$ and 30.14°. These peaks are characteristic of monetite phase ($CaHPO_4$) according to 01-077-0128 reference from the ICDD-PDF22003 database. The increase of the amount of biopolymer induced the broadening of peaks, which proves their incorporation on the apatitic surface.

4.1.2 FT-IR spectroscopy

Figures 4 and **5** shows the FT-IR spectra of CaHAp, CaHAp-Alg and CaHAp-(Carr). In all IR spectra, the vibration bands of PO_4^{3-} groups of the apatite structure are observed at (ν_s) 965 cm^{-1}, (δ_s) 482 cm^{-1}, (ν_{as}) 1041-1094 cm^{-1} and (δ_{as}), 567-605 cm^{-1} [25]. Moreover, for characteristic bands of hydroxyl ions were observed towards (ν_s) 3572 cm^{-1} and (ν_L) 634 cm^{-1} [26].

The two broad bands located at 1403-1462 and 1640 cm^{-1} were assigned, respectively, to the carbonate ions and water adsorbed on the surface. Besides, the band located at 877 cm^{-1} was assigned to the (HPO_4^{2-}) group [27].

After grafting, the intensity of the hydroxyls bands (υ_s and ν_L) decrease progressively with increasing Alg or (λ-Carr) amount. This can be explained by the low degree of crystallinity [28].

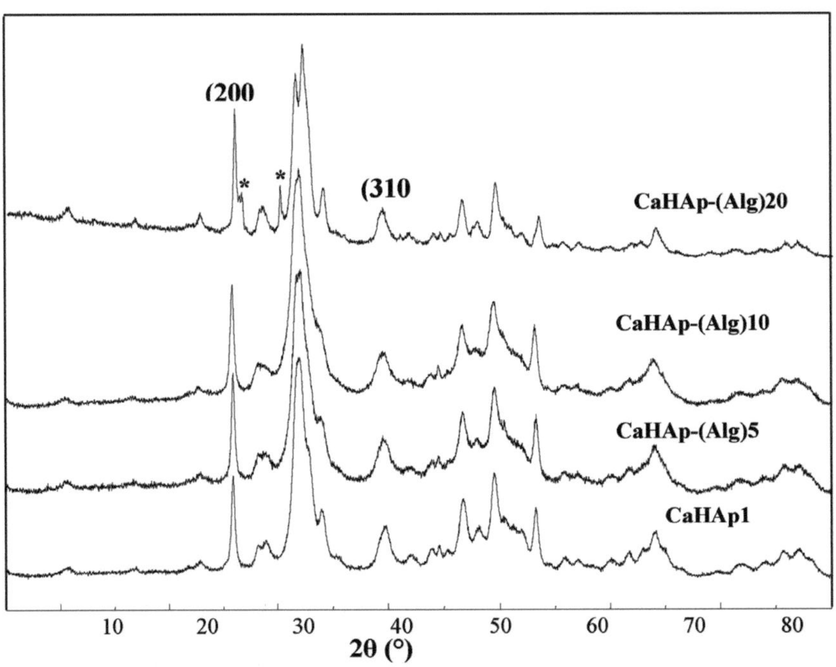

Figure 2.
Powder x-ray diffraction patterns of modified hydroxyapatites by sodium alginate.

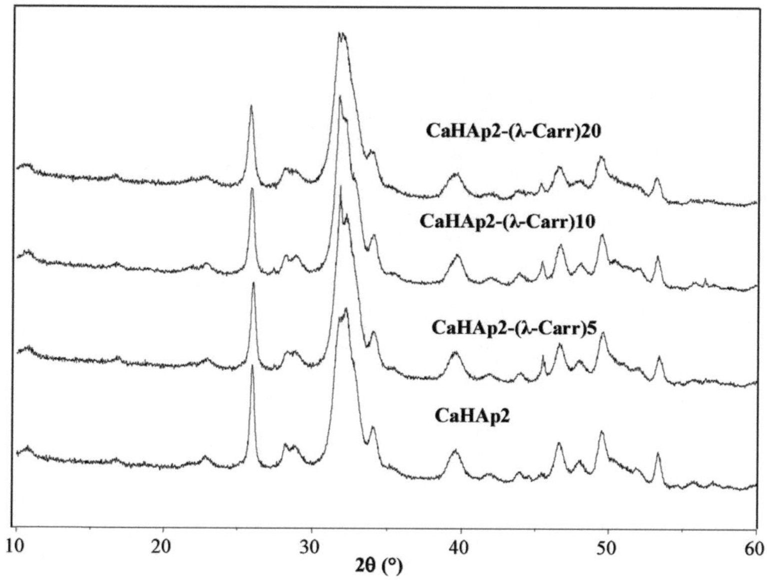

Figure 3.
Powder x-ray diffraction patterns of modified hydroxyapatites by lambda carrageenan.

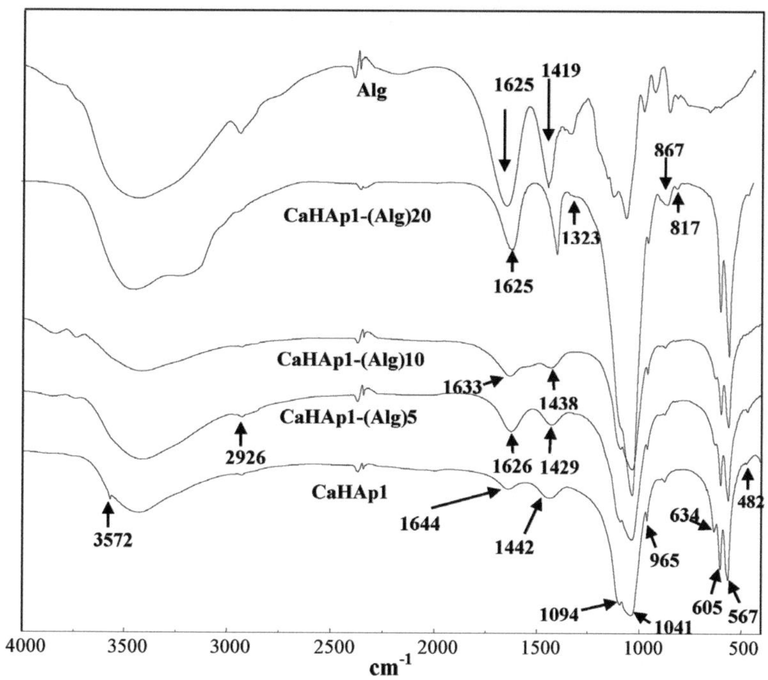

Figure 4.
FTIR spectra of modified hydroxyapatites by sodium alginate.

The presence of biopolymers (Alg or carr) on the apatitic surface is confirmed by the appears of new vibrations bands at 1231 cm^{-1} and (1626, 1323 and 1428 cm^{-1}) which are assigned respectively, to S-O groups of (λ-Carr) and COO^- groups of (Alg) [29, 30].

On the other hand, the conservation in the band intensity of P-OH group at 870 cm^{-1} indicates that fixing process of (λ-Carr) or Alg is done only by the interaction

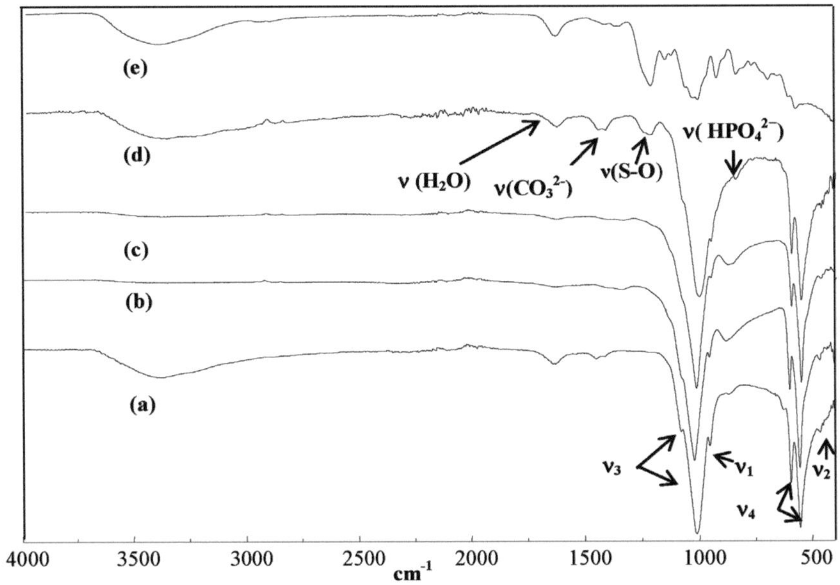

Figure 5.
FTIR spectra of modified hydroxyapatites by lambda carrageenan.

between ≡Ca-OH groups of apatitic surface and functional groups of each biopolymer.

The possible modes of interaction of (Alg) or (λ-Carr) with the CaHAp1 or CaHap2 surface have been gathered in **Figures 6** and 7.

4.1.3 Scanning electron microscope (SEM)

Figure 8 shows the morphological variations of CaHAp1 or CaHAp1 before and after reaction with Alg or (λ-Carr). The CaHAp1 or caHap2 is composed of irregular particles with a strong tendency to aggregate (**Figure 8a, b**), whereas the modified hydroxyapatites showed that their shapes become relatively agglomerates

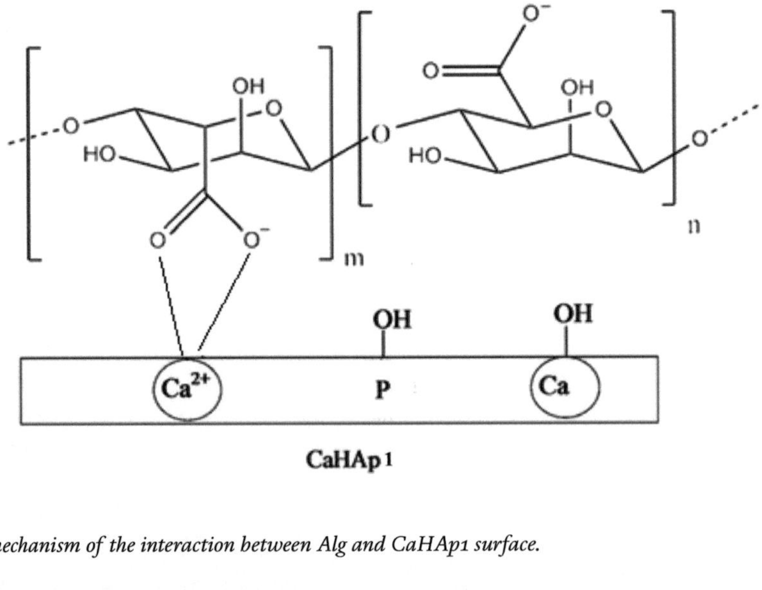

Figure 6.
Proposed mechanism of the interaction between Alg and CaHAp1 surface.

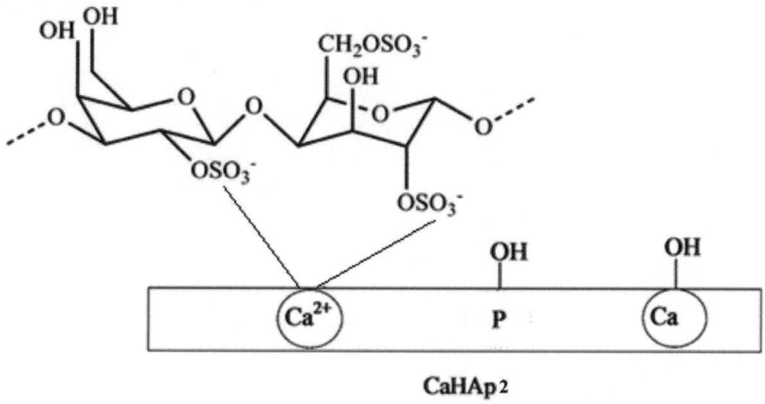

Figure 7.
Proposed mechanism of the interaction between (λ-Carr) and CaHAp2 surface.

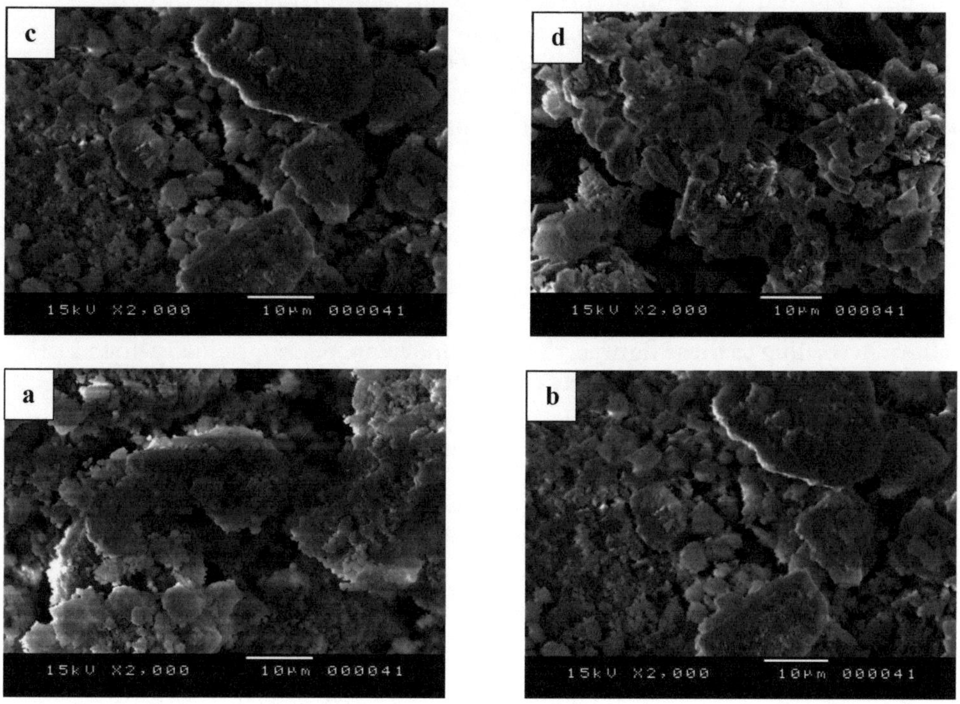

Figure 8.
SEM photomicrograph of modified hydroxyapatites: (a) CaHAp1, (b) CaHAp2, (c) CaHAp1-(Alg)10 and (d) CaHAp2-(Carr)10.

of different sizes and poorly defined shape (**Figure 8(c, d)**). This change is due to the formation of new hybrid compounds CaHAp2–(λ-Carr) and CaHAp1-(Alg).

4.1.4 Textural properties

The surface characteristics of the Hydroxyapatite samples obtained by the BET method, both before and after modification, are shown in **Table 1**. The treatment of hydroxyapatite with Alg biopolymer leads to the decreasing of specific surface area compared to that of ungrafted CaHAp1, covering the range from 20.52 to 3.39 m^2/g. According to BET results, The fact that the surface area of CaHAp1-(Alg)5 and

Samples	Surface area (m^2/g)
CaHAp1	21.64
CaHAp2	93.00
CaHAp1-(Alg)5	12.84
CaHAp1-(Alg)10	3.39
CaHAp2-(λ-Carr)5	168.00
CaHAp2-(λ-Carr)10	253.00
CaHAp2-(λ-Carr)20	260.00

Table 1.
Measurement of surface area of grafted hydroxyapatites.

CaHAp1-(Alg)10 materials is lower than those in CaHAp1 can be explained with the filling of pores on the surface of hydroxyapatite by sodium alginate biopolymer. On the other hand, the specific surface of modified hydroxyapatite by lambda carrageenan increases with the increasing of grafting rate. The maximum value is obtained for CaHAp2-(λ-Carr) 20 (SSA = 260 m^2/g).

4.2 Evaluation of the performance of the prepared compounds for the adsorption of MB

4.2.1 Effect of pH

Figures 9 and **10** represent the variation of the adsorption capacity of MB on the surface of CaHAp1, CaHAp2, CaHAp1-(Alg) and CaHAp2-(λ-Carr) at different pH values. According to these figures, the MB removal increases gradually up to a pH very close to 5, and after kept nearly constant with further pH increase. The effect

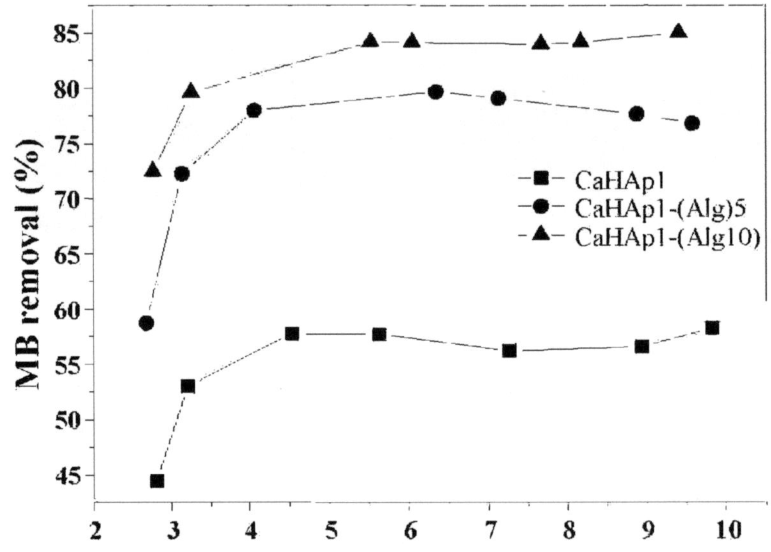

Figure 9.
Effect of initial pH on the adsorption of MB (Adsorbent dosage = 1 g/L, initial MB concentration = 50 mg/L, solution volume = 50 mL, temperature = 25°C, contact time = 3h).

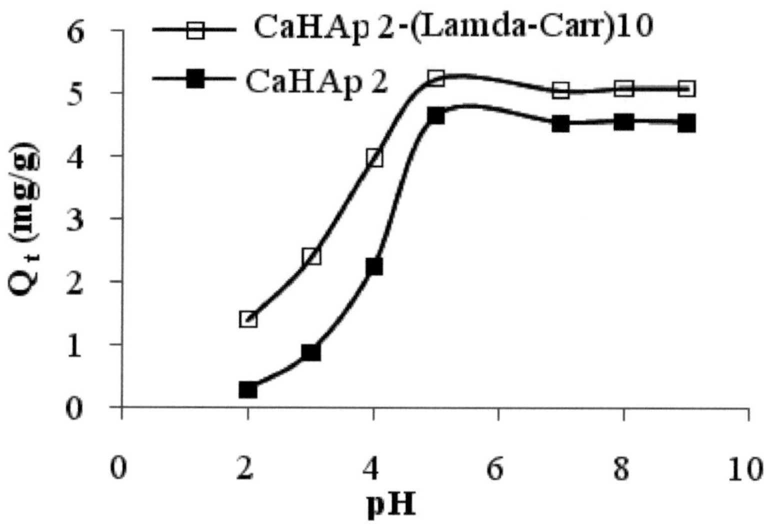

Figure 10.
Effect of initial pH on the adsorption of MB (Adsorbent dosage = 1 g/L, initial MB concentration = 10 mg/L, solution volume =25 mL, temperature = 25°C, contact time = 3h).

of pH on adsorption depends on the point zero charge (pH pzc) of adsorbent. The value of pHpzc was found to be 6.2, 6.7 and 7.3 by CaHAp1, CaHAp2-(Carr)10 and CaHAp1-(Alg)10, respectively, in good agreement with literature data [31]. Indeed, for pH< pH pzc, the surface of hydroxyapatite is positively charged which causes the repulsion of cationic groups of the MB molecule (=N^+), and hence low adsorption of dye. In the case of modified hydroxyapatite, the amount of dye adsorbed is higher than unmodified because (Alg) or (λ-Carr) contains carboxylate (–COO^-) and sulphonate (OSO_3^-) groups, respectively, that increase the interaction with dye molecules.

4.2.2 Effect of contact time and adsorbent dose

The adsorptions of MB onto modified hydroxyapatites were studied with different biopolymer doses and at different periods of time. As shown in **Figures 11** and **12**, the adsorption amounts of MB on the modified hydroxyapatites increased rapidly with an increase in time and then became slower until the equilibrium was reached. Equilibrium is reached for all compounds prepared at low durations between 20 and 30 min. This may be due to the rapid saturation of the pores of ours prepared adsorbents. The maximum adsorption is obtained for hydroxyapatite synthetized in the presence of biopolymer with 10% content. Therefore, the increase of MB removal with the increasing of adsorbent dose is mainly due to the increasing of interaction forces between the carboxylate groups (COO^-) of (Alg) or sulphonate goups (OSO_3^-) of (λ-Carr) and MB molecules via N^+ groups. The possible modes of interaction between MB and either CaHAp1-(Alg) or or CaHAp2-(λ-Carr) surface are given in **Figures 13** and **14**.

4.2.3 Effect of temperature and initial dye concentration on the adsorption process

Figures 15 and **16** shows the effect of temperature on the adsorption of MB on the surface of the prepared supports CaHAp1, CaHAp2, CaHAp1-(Alg) and CaHAp2-(λ-Carr). As it is observed, the removal of MB by CaHAp2 increases with increasing the temperature of the solution from 25 to 60°C, indicating that

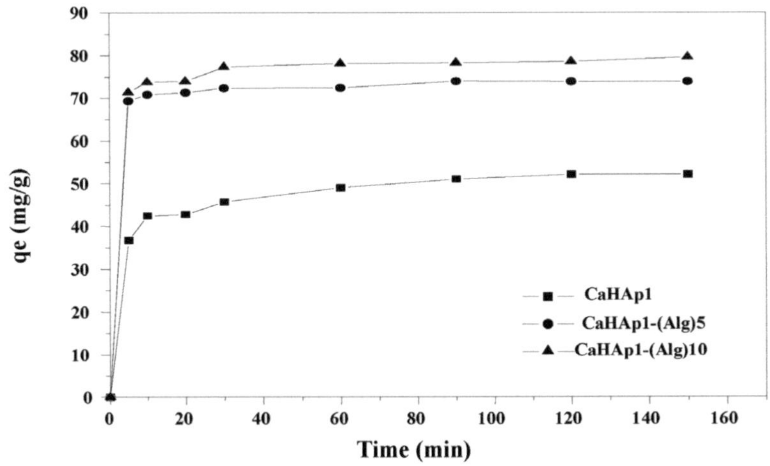

Figure 11.
Effect of contact time on MB adsorption by CaHAp1 and CaHAp1–Alg (adsorption conditions: 0.05 g of adsorbent, initial MB concentration = 50 mg/L, solution volume = 50 mL, 6 < pH< 7, rpn =110 tr/min and temperature = 25 °C).

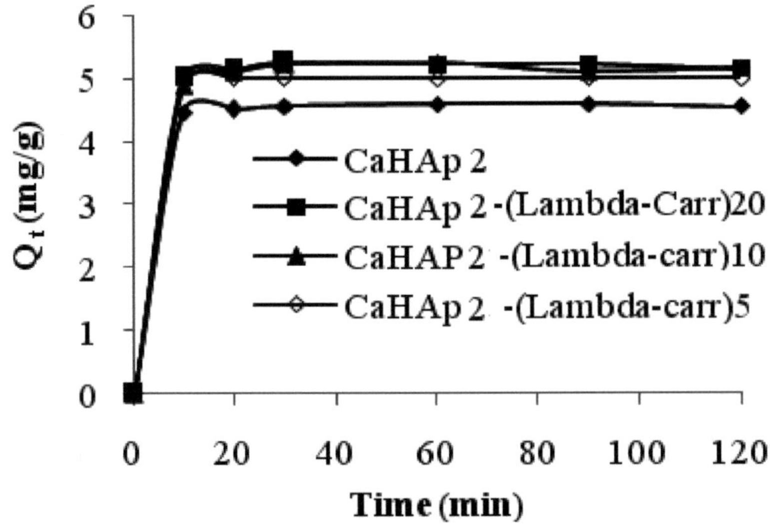

Figure 12.
Effect of contact time on MB adsorption by CaHAp1 and CaHAp2–(λ-Carr) (adsorption conditions: 0.05 g of adsorbent, initial MB concentration = 50 mg/L, solution volume = 50 mL, 6 < pH< 7, rpn =110 tr/min and temperature = 25 °C).

the process is endothermic. In the case for other absorbents, the decrease of the adsorption of MB with the increase of temperature indicates that the adsorption is exothermic in nature.

This phenomenon could be explained by the decrease of the interaction between the MB ions and active sites in hydroxyapatite surface and the weakening of adsorptive forces between the carboxylate (COO^-) groups of (Alg) or sulphate (OSO_3^-) groups of (λ-Carr) and cationic dye molecules. As also depicted from **Figures 15** and **16**, the adsorbed amounts increase with the initial dye concentration to a threshold concentration corresponding to the saturation of the adsorption sites.

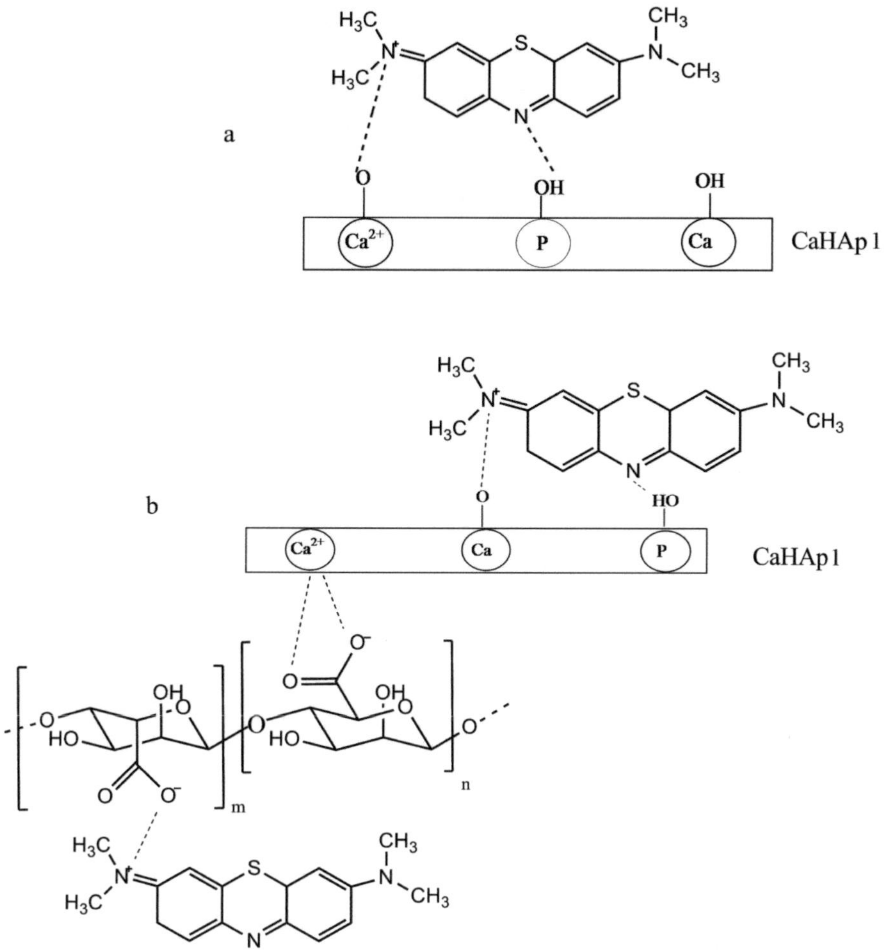

Figure 13.
The possible modes of interaction between MB and either (a) CaHAp1 or (b) CaHAp1-(Alg).

According to these figures, the maximum adsorption capacity of CaHap1-(ALg)10 is more important than those of CaHAp2-(λ-Carr)10, CaHAp1 and CaHAp2. The adsorbed quantities are respectively 128.4, 98.23, 68.5 and 58.8 mg/g. Compared to other adsorbents gathered from the literature (**Table 2**), this registered amount of dye removal is so very interesting and thus our developed product could be seen as a good adsorbent.

In fact, this value is more important com-pared to garlic peel used as adsorbent (82.64 mg/g), Raw date pits (80.3 mg/g) and wood (84 mg/g). It is nearly twice higher compared to Wood ash (50 mg/g). It is five times more important than the Cotton waste (24 mg/g) and. It is six times important than modified pumice stone (15.87 mg/g) and Pure kaolin (15.55 mg/g). It is hundred times important than Fly ash (1.3 mg/g). Consequently, the above results confirmed that CaHAp-(Alg)10 was a favorable adsorbent for MB dye removal.

4.2.4 Modeling and determination of kinetic parameters

In order to evaluate the kinetics involved in the process of the adsorption of MB dye onto CaHAp1, CaHAp2, CaHAp2-(Carr)10 and CaHAp1-(Alg)10 surface,

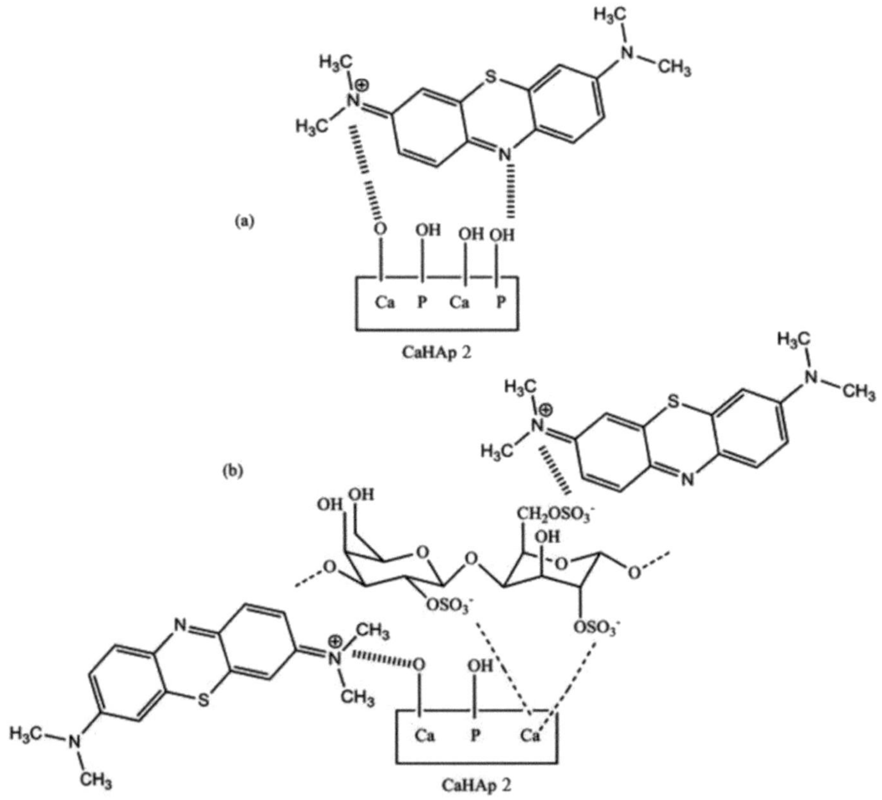

Figure 14.
The possible modes of interaction between MB and either (a) CaHAp2 or (b) CaHAp2-(λ-Carr).

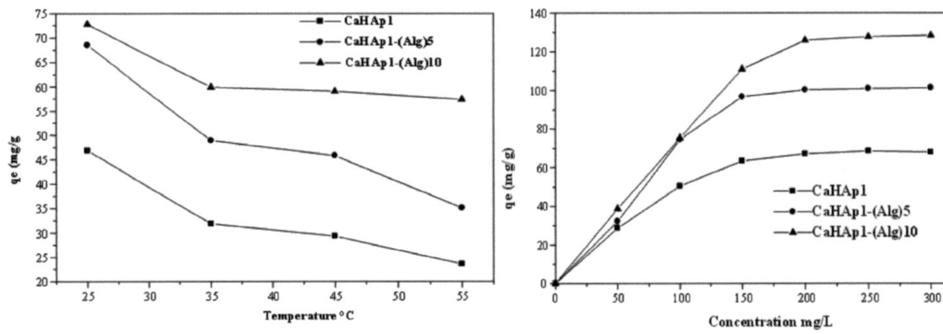

Figure 15.
Effect of temperature (a) and initial concentration (b) on the removal of MB using CaHAp1 and CaHAp1-(Alg).

pseudo-first-order, pseudo-second-order, Elovich, and intra-particle diffusion models were applied and results were discussed [42].

The results of the kinetic study, for hydroxyapatite synthesized in the presence of different amounts of Alg or (λ-Carr) are summarized in **Tables 3** and **4**. The best fit model was selected based on the linear regression correlation coefficient, R^2 values. According to the regression coefficients ($R^2 > 0.99$), the pseudo-second-order equation appears more suitable to describe the retention of MB on all prepared supports.

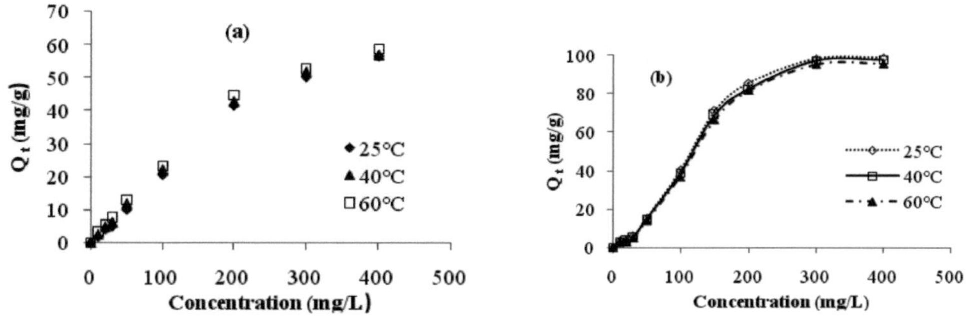

Figure 16.
Effect of temperature on the removal of MB using: (a) CaHAp2 and (b) CaHAp2-(λ-Carr)10.

Samples	q_m (mg/g)	References
CaHAp1-(Alg)10	128.40	In this work
Garlic peel	82.64	[32]
Wood ashes	50	[33]
Cotton waste	24	[34]
Modified pumice stone	15.87	[35]
Natural phosphate	7.23	[36]
natural zeolite	16.370	[37]
Raw date pits	80.30	[38]
Pure kaolin	15.55	[39]
Fly ash	1.3	[40]
Wood	84	[41]

Table 2.
Comparison of the maximum BM adsorption capacity of the adsorbents in this study with other adsorbents.

Consequently, it can be confirmed that the adsorption is chemical and assumes that the surface of our adsorbent is heterogeneous.

4.2.5 Adsorption isotherm

In this study, Langmuir, Freundlich, Tempkin and Dubinin–Radushkevich (D-R) isotherm models were used to describe the adsorption of MB onto ungrafted and grafted hydroxyapatites [43].

The obtained parameters of each model are given in **Tables 5** and **6**. According to the analysis of regression coefficients (R^2), Langmuir model has the highest value than those of the other models for modified hydroxyapatite by sodium alginate. This indicates that the Langmuir model appears appropriate for modeling the adsorption isotherms of MB on CaHAp1 or CaHAp1-(Alg) surface. As seen in **Table 5**, the R_L values [44] are situated within the range of $0 < RL < 1$, which means that prepared CaHAp or CaHAp-(Alg) is favorable for the adsorption of MB dye under the experimental conditions conducted in this study.

For modified hydroxyapatite by lambda carrageenan, the analysis of regression coefficients shows better applicability of Freundlich model. Moreover, in all compounds, the values of calculated n are less than 1, indicating that the adsorption of MB is favorable.

	Pseudo-first-order			Pseudo-second-order			Elovich			Intra-particular diffusion	
	K_1	q_e	R^2	K_2	q_e	R^2	α	β	R^2	K_i	R^2
0%	0.0234	16.764	0.8305	0.00648	52.910	0.9991	12.946	12.210	0.8131	3.0578	0.6086
5%	0.149	4.769	0.3244	0.02641	74.074	0.9999	28.701	11.128	0.5838	3.505	0.3675
10%	0.0181	7.197	0.4388	0.01556	79.365	0.9999	28.931	8.903	0.6240	3.9032	0.4046

Table 3.
Kinetic constants for the adsorption of studied dyes on hydroxyapatite-Alginate (T = 25°C, C_0 = 100 mg/L, 6 < pH < 7).

	Pseudo-first-order			Pseudo-second-order			Elovich			Intra-particular diffusion	
	K_1	q_e	R^2	K_2	q_e	R^2	α	β	R^2	K_i	R^2
0%	0.0088	0.41	0.346	2.7	4.58	0.999	7.49^{E}18	10.41	0.670	0.317	0.488
5%	0.0085	0.29	0.271	5.51	5.02	1	1.33^{E}57	27.24	0.457	0.344	0.476
10%	0.0075	0.53	0.247	−20.86	5.162	0.999	1.31^{E}10	5.17	0.378	0.361	0.498
20%	0.0079	0.34	0.2	−1.18	5.184	0.999	3.67^{E}15	7.62	0.349	0.361	0.489

Table 4.
Kinetic constants for the adsorption of studied dyes on hydroxyapatite-Carrageenan ($T = 25°C$, $C_o = 10$ mg/L, pH = 5).

Isotherm model	Adsorption constant	Adsorbent		
		CaHAp1	CaHAp1-(Alg)5	CaHAp1-(Alg)10
Langmuir	q max (mg/g)	77.51	117.64	142.85
	K_L (l/mg)	0.038	0.041	0.062
	R_L	0.34-0.08	0.33-0.07	0.25-0.04
	R^2	0.994	0.974	0.992
Freundlich	n	2.876	0.468	0.541
	K_F (l/mg)	11.762	0.0028	0.0209
	R^2	0.864	0.827	0.685
Tempkin	B_1	18.462	0.032	0.028
	K_T	0.286	$3.8 \cdot 10^{19}$	$4.6 \cdot 10^{13}$
	R^2	0.946	0.777	0.886
D-R	q max (mg/g)	66.62	105.50	120.50
	$K_{D\text{-}R}$ (mol²/J²)	$6 \cdot 10^{-5}$	$5 \cdot 10^{-5}$	$2 \cdot 10^{-5}$
	E (KJ/mol)	91.28	100	158.11
	R^2	0.97	0.95	0.94

Table 5.
Isotherm constants and correlation coefficients (R^2) for MB adsorption on the surface of CaHAp1 and CaHAp1-(Alg) at different temperatures.

The values of E calculated from Dubinin-Redushkevich equation are greater than 40 KJ mol^{-1} for all samples, which confirms that the adsorption is chemical [42].

4.2.6 Determination of the thermodynamic parameters

The standard Gibbs free energy change (ΔG°) has been determined from the following equation [21].

$$\Delta G° = -RT\,LnK_L \tag{4}$$

Where R is the universal gas constant (8.314 J mol^{-1} K^{-1}), T is the temperature (K) and K_L value is the Langmuir equilibrium constant. The enthalpy change ΔH°

	Langmuir constants			Freundlich constants			Temkin constants			Dubinin-redushkevich constants			Thermodynamic parameters		
	CaHAp2														
T (°C)	K_L	$q_{m.L}$	R^2	K_F	n_F	R^2	B_T	A_T	R^2	$Q_{m.DR}$	E	R^2	ΔH°	ΔS°	ΔG°
25	0.017	285.7	0.407	0.51	1.03	0.990	36.85	0.29	0.916	3.84	70.71	0.596	42.3	0.25	−32.2
40	0.025	156.2	0.851	0.62	1.12	0.991	37.1	0.30	0.925	3.98	70.71	0.588			−26.95
60	0.026	125	0.944	0.73	1.21	0.993	37.52	0.31	0.928	4.1	79.05	0.572			−40.95
	CaHAp2-(λ-Carr)10														
25	0.005	588.2	0.161	0.894	0.951	71.75	0.289	0.909	4.9	70.71	0.483	0.328	−3.1	−29.2	8.69
40	0.005	526.3	0.189	0.89	0.953	70.52	0.288	0.909	4.85	70.71	0.481	0.323			9.13
60	0.004	500	0.181	0.87	0.955	69.32	0.286	0.905	4.77	70.71	0.486	0.314			9.7

Table 6.
Isotherm constants and thermodynamic parameters for dye adsorption on the surface of CaHAp2 and CaHAp2-(Carr)10 at different temperatures.

	$\Delta H°$ (KJ/mol)	$\Delta S°$ (KJ/mol)	$\Delta G°$ (KJ/mol)			
			298°K	308°K	318°K	328°K
CaHAp1	−27.96	−0.094	0.218	1.163	2.109	3.055
CaHAp1-Alg(10)	−20.432	−0.059	−2.789	−2.197	−1.605	−1.013

Table 7.
Values of thermodynamic parameters for MB dye removal with CaHAp1 and CaHAp1-Alg.

and entropy change $\Delta S°$ of the adsorption were estimated from the following equation:

$$LnK_L = -\frac{\Delta H^0}{RT} + \frac{\Delta S^0}{R} \quad (5)$$

The values of $\Delta H°$ and $\Delta S°$ were determined from the slopes and intercept of the linear plot of Ln K_L versus (1/T). The calculated thermodynamic parameters are summarized in **Tables 6** and 7.

The values of the enthalpy $\Delta H°$ indicates that the adsorption process is endothermic for CaHAp2 ($\Delta H° > 0$), and exothermic for (CaHAp1, CaHap2-(λ-Carr)10, CaHAp1-(Alg)10) ($\Delta H° < 0$). The $\Delta G°$ values were negative for (CaHap2, CaHAp1-(Alg)10) and positive for (CaHAp1, CaHAp2-(λ-Carr)10).

The positive $\Delta G°$ values indicates that the instability activation complex of the adsorption reaction increases with increasing temperature, The negative values of $\Delta G°$ means that the process is feasible and adsorption is spontaneous thermodynamically [45].

The negative value of entropy $\Delta S°$ confirms the decreased randomness at the solid–solution interface during adsorption.

5. Conclusion

In this study, the results of characterization techniques IR, XRD, SSA and SEM analysis showed that the sodium alginate or lambda carrageenan were successfully grafted on the hydroxyapatite surface. The modified hydroxyapatite could be potentially applied for the removal of methylene blue dye from aqueous solution. The determination of the amount of dye adsorbed on the various apatitic phase CaHAp1, CaHAp2, CaHAp1-(Alg) and CaHAp2-(λ-Carr) allowed us to build the adsorption isotherms which give information on the adsorption mechanism. The Modeling of the isotherms proved that the adsorption of MB on modified hydroxyapatite is described by the Freundlich equation for CaHAp2-(λ-Carr) and Langmuir for CaHAp1-(Alg). The calculated adsorption capacities of CaHAp1, CaHAp2, CaHAp1-Alg10 and CaHAp2-(λ-Carr)10 for MB, at 25°C, were 68.5, 58.8 128.40, and 98.23 mg/g, respectively. The results gave a clear indication that the modified hydroxyapatite has a great capacity than pure hydroxyapatite for dye removal.

Thermodynamic studies indicated that the physic-sorption is the dominating mechanism for the dye adsorption process onto CaHAp1-(Alg)10 or CaHAp2-(λ-Carr)10, spontaneous and exothermic in nature. The results indicate that the modified hydroxyapatite CaHAp2-(λ Carr) or CaHAp1-(Alg) possessed good adsorption ability towards MB dye and can be used as a low cost adsorbent for other environmental applications as retention of heavy metal and pesticides from wastewater. Further works will be extended for the functionalization of hydroxyapatite materials with surfactants and cationic reagents for the removal of organic pollutants from contaminated waters.

Acknowledgements

The authors thank the INRAP (Institut Nationale de Recherche et d'Analyses Physico-chimiques), INRST (Institut Nationale de Recherche Scientifique et Technologique), ETAP (Entreprise Tunisienne d'Activités Pétrolières) and the University of El Manar and Monastir (Tunisia).

Author details

Hassen Agougui[1,2*], Youssef Guesmi[3,4] and Mahjoub Jabli[5,6]

1 Faculty of Sciences of Gafsa, University of Gafsa, University Campus, Zarroug, Gafsa, Tunisia

2 Laboratory of Physical-Chemistry of Materials, Faculty of Sciences of Monastir, Monastir, Tunisia

3 Laboratory, Water, Membrane and Environmental Biotechnology, Centre of Research and Water Technologies, Technopark of Borj-Cedria, Soliman, Tunisia

4 Faculty of Sciences of Tunis, University of Tunis - El Manar, Tunis, Tunisia

5 Department of Chemistry, College of Science Al-zulfi, Majmaah University, Al-Majmaah, Saudi Arabia

6 Textile Materials and Processes Research Unit, Tunisia National Engineering School of Monastir, University of Monastir, Tunisia

*Address all correspondence to: hassenagougui@yahoo.fr

References

[1] D. Suteu, S. Coseri, C. Zaharia, G. Biliuta, I. Nebunu, *Desalination Water Treat*. 90 (2017) 341-349

[2] G. Annaduraia, R.S. Juang, D.J. Lee, *J. Hazard. Mater*.92 (2002) 263-274

[3] S. A. Shahid Chatha, M. Asgher, S. Ali, A. Ijaz Hussain, *Carbohydr. Polym*. 87 (2012) 1476-1481.

[4] M. R. Fat'hi, A. Ahmadi, *Int J Env Health Eng*. (2016) 5-19

[5] S. Masoudnia, M. H. Juybari, R. Zafar Mehrabian, M. Ebadi, F. Kaveh, *Int. J. Biol. Macromol*. 165 (2020) 118-130.

[6] G. J. Copello, A. M. Mebert, M. Raineri, M. P. Pesenti, L. E. Diaza, *J. Hazard. Mater*. 186 (2011) 932-939.

[7] Y. Wang, Y. Peib, W. Xiong, T. Liud, J. Li, S. Liu, B. Li, *Int. J. Biol. Macromol*. 81 (2015) 477-482.

[8] N. A. Zakharov, Zh. A. Ezhova, E. M. Koval, V. T. Kalinnikov, and A. E. Chalykh, *Inorg. Mate*r. 41 (2005) 509-515.

[9] R. Khanna, K. S. Katti, and D. R. Katti, *J. Polym. Sci*. 2010 (2010) 1-12.

[10] D. Wahl and J. Czernuszka, *Eur Cell Mater*. 11 (2006) 43-56.

[11] K. Senthilarasan, A. Ragu and P. Sakthivel, *Int. J. Eng. Res. Appl*. 4 (2014)55-59.

[12] N. Koupaei, A. Karkhaneh, J. Tissue. *Eng. Regen. Med*. 13 (2016) 251-260.

[13] K. Kanimozhi, D. Gopi and L. Kavitha, *Int. J. Eng. Sci. Eng. Res*. 5 (2014) 138-140.

[14] H. Hou, R. Zhou, P. Wu, L. Wu, *Chem. Eng. J*. 211-212 (2012) 336-342.

[15] M. C. Chang, C. Chang Ko, W. H. Douglas, *Biomaterials*. 24 (2003) 2853-2862.

[16] M. Vila, S. Sánchez-Salcedo, M. Cicuéndez, I. Izquierdo-Barba, María Vallet-Regí, *J. Hazard. Mater*. 192 (2011) 71-77.

[17] K. Sangeetha, G. Vasugi, E. K. Girija, *Int. J. Chem. Tech. Res*. 8 (2015) 117-125.

[18] G. Vasugi and E. K. Girija, *Chem. Technol*. 49 (2015) 87-91.

[19] W. Wei, R. Sun, Z. Jin, J. Cui, Z. Wei, *Appl. Surf. Sci*. 292 (2014) 1020-1029.

[20] U. Nadeem and M. Datta, *Eur. Chem. Bull*. 3 (2014) 682-691.

[21] H. Hou, R. Zhou, P. Wu, L. Wu, *Chem. Eng. J*. 211-212 (2012) 336-342.

[22] A. K. Radzimska, M. Samuel, D. Paukszta, A. Piaseck, T. Jesionowsk, *Physicochem. Probl. Miner. Process*. 50 (2014) 225–236.

[23] H. Eslami, M. S. Hashjin, M. Tahriri, Iranian *J. Pharm. Sci*. 4 (2008) 127-134.

[24] K. Allam, A. El Bouari, B. Belhorma, L. Bih, *J.WA.R.P*. 8 (2016) 358-371.

[25] A. Aissa, H. Agougui, M. Debbabi, *Appl. Surf. Sci*. 257 (2011) 9002-9007.

[26] H. Agougui, A. Aissa, S. Maggi, M. Debbabi, *Appl. Surf. Sci*. 257 (2010) 1377.

[27] M. Othmani, A. Aissa, H. Bachoua, M. Debbabi, *Appl. Surf. Sci*. 264 (2013) 886-891.

[28] L. Sukhodub, *Mater. Sci. Technol*. 40 (2009) 318-325.

[29] Malik, Z. M.; Jarmoluk, K. M. A. *Polymer*. 8 (2016) 275.

[30] N.A. Kamalaldin, B.H. Yahya, A. Nurazreena, *Procedia. Chem*. 19 (2016) 297-303.

[31] Wei, W.; Yang, L.; Zhong, W. H.; Li, S. Y.; Cui, J.; Wei, Z. G. Dig. *J. Nanomater. Biostruct*. 10 (2015) 1343.

[32] B.H. Hameed, A.A. Ahmad, *J. Hazard. Mater*. 164 (2009) 870-875.

[33] N. E. Fayoud, S. A. Younssi, S. Tahiri, A. A. Albizane, *J. Mater. Environ. Sci*. 6 (2015) 3295-3306.

[34] G. McKAY, G. RAmprasad, P. P.RATAPA MOWLI, *Water, Air, and Soil Pollution*. 29 (1986) 273-283.

[35] Z. Derakhshan, M. A. Baghapour, M. Ranjbar, M. Faramarzian, *Health Scope*. 2 (2013) 136-144.

[36] N. Barka, A. Assabbane, A. Nounah, L. Laanab, Y.A. Ichou, *Desalination*. 235 (2009) 264-275.

[37] Runping Han,, Jingjing Zhang, Pan Han, Yuanfeng Wang, Zhenhui Zhao, Mingsheng Tang, *Chem. Eng. J*. 145 (2009) 496-504.

[38] F. Banat, S. Al-Asheh, L. Al-Makhadmeh, *Process Biochem*. 39 (2003) 193-202.

[39] D. Gosh, G. Bhattacharyya, *Appl. Clay. Sci*. 20 (2002) 295-300.

[40] C. Woolard, J. Strong, C. Erasmus, *Appl Geochem*. 17 (2002) 1159-1164.

[41] G. McKay, V. Poots, *J. Chem. Technol. Biotechnol*. 30 (1986) 279-282

[42] Runping, H.; Pan, H.; Zhaohui, C.; Zhenhui, Z.; Mingsheng, T. *J. Environ. Sci*. 2008, 20, 1035.

[43] W. Wei, L. Yang, W.H. Zhong, S.Y. Li, J. Cui, Z.G, Wei, Dig. *J. Nanomater. Biostruct*. 10 (2015) 1343-1363.

[44] D. Robati, B. Mirza, R. Ghazisaeidi, M. Rajabi, O. Moradi, S. Agarwal, V.K. Gupta, I. Tyagi, *J. Mol. Liq*. 216 (2016) 830-835.

[45] Z. Jia, Z. Li, S. Li, Y. Li, R. Zhu, *J. Mol. Liq*. 220 (2016) 56-62.

Chapter 11

Treatment of Textile Dyeing Waste Water Using TiO_2/Zn Electrode by Spray Pyrolysis in Electrocoagulation Process

Parameswari Kalivel

Abstract

An alternative form of treatment for the remediation of textile waste water, electrocoagulation (EC) methods are used. This work deals mainly with the treatment of waste water for textile dyeing preceded by the use of wastewater. The goal of the proposed study is to evaluate the efficiency of the electrocoagulation process using TiO_2/Zn electrodes using $TiCl_3$ via spray pyrolysis. The surface morphology of the electrode was studied by SEM, XRD and EDS analysis. The efficiency of electrocoagulation treatment process to treat synthetic waste water containing Coralene Navy RDRLSR, Coralene Red 3G, Rubru RD GLFI dye was studied. The effect of parameters such as current density, influence of effluent pH, supporting electrolyte NaCl concentration, and EC time on dye removal efficiency were investigated. The result indicates that this process is very efficient and was able to achieve color removal (99.5%) at pH 8.5 and 0.15 A in 10 minutes.

Keywords: electrocoagulation, textile dye effluent, TiO_2/Zn electrode, EC time

1. Introduction

1.1 Water crisis

Water is the source of life and regarded as the most essential resource of natural resources. In recent days, demand for water has increased due to tremendous growth in technology and industries. The ever increasing world population and rapidly advancing industrialization is causing more demand than ever for the dwindling supply of water, which makes it precious [1]. Many water pollutants are yet to be addressed, due to rapid industrialization there are new pollutants which are being discovered. New chemical compounds are continuously being developed and brought to the market and sooner or later they will be utilized in the aquatic systems. There may be adverse effects on human health and marine environments from emerging contaminants found in water. For human health, safe water that is free of harmful chemicals and pathogens is essential [2]. Water shortages will contribute to social and political unrest, water wars and diseases in the coming decades, and it will take a dramatic turn in the opposite direction unless new methods of providing clean water are found. Increased public awareness has led governments and organizations worldwide to issue strict water pollution regulations. These days there is an increasing worldwide concern for the development of wastewater treatment technology [3]. The conversion of waste water into reusable water can be done by using scientific methods and that decreases the challenge towards the rising demand of water quality [4]. Comparing all industries, textile industries show a higher rate of consumption of water.

1.2 Status of pollution due to textile industries

Dyes can be characterized as substances that when applied to a substrate, provide color through a process that alters the crystal structure of the colored substances, at least temporarily. In the textile, pharmaceutical, food, cosmetics, plastics, photographic and paper industries, such substances with substantial coloring capability are commonly used. Dyes are categorized by application and chemical structure and consist of a group of atoms known as chromophores that are responsible for the color of the dye. The numerous functional groups, such as azo, anthraquinone, methine, nitro, carbonyl and others are used as chromophore-containing centers in dyes, in addition, electrons removing or donating substituents are called auxochromes in order to produce or enhance the color of the chromophores are also present in dyes. Over 10,000 different dyes and pigments are expected to be used industrially and over 7×10^5 tons of synthetic dyes are manufactured globally annually [5–7]. Using batch, continuous or semi-continuous processes, textile products can be dyed. The type of method used depends on many features, including the type of material such as fibre, yarn, fabric, construction of garments and clothes, as well as the generic type of fibre, the size of dye batches and the quality specifications of the dyed fabric. The batch process is the most common method of dyeing textile materials among these methods. In the textile industry, due to the inefficiency of the dyeing process, up to 200,000 tons of these dyes are lost to effluent every year during the dyeing and finishing operations. Unfortunately, as a result of their high stability to light, temperature, water, detergents, chemicals, soap and other parameters such as bleach and transpiration, most of these dyes avoid traditional wastewater treatment systems and remain in the setting. Furthermore, in the manufacture of textiles, anti-microbial agents resistant to biological degradation are frequently used, especially for natural fibres such as cotton. They are more recalcitrant to biodegradation by the synthetic origin and complex aromatic structure of these agents. A significant volume of water used mainly in the dyeing and finishing operations of the plants is used in the textile industry in its production methods. Both dry and wet processing include a textile unit. In an extended process series that produces a large amount of waste, the textile manufacturing process is distinguished by the high use of resources such as water, fuel and a broad variety of chemicals. A huge amount of solid and liquid waste is produced by textile processing units, some of which may be hazardous [8]. Considering the volume produced as well as the composition of the effluent, waste water from textile plants is listed as the most polluting of all industrial sectors. Moreover the increased demand for textile goods and the proportional increase in their production, along with the use of synthetic dyes, have led to make dye wastewater one of the key sources of serious pollution problems at the present time.

The nature of the waste produced depends on the range of fibres and chemical substances involved in the textile process, the processing methods and technologies that are being adopted. Usually the most important ecological problems associated with textile production are those associated with the degradation of the body of water caused by the discharge of raw waste [9]. Dyeing and finishing units are a key part of the produced waste water as they use a high volume of water in their operation, from fibre washing to bleaching, dyeing and washing of refined products [10]. On average, there is a sample mixture of compounds in around 200 L of water that could cause harm if strict precautions are not taken until they are released into the atmosphere. Wet process manufacturing utilizes color caustic soda, sulphuric acid, sodium peroxide, hydrochloric acid, dyes and chemicals. Moreover, other metals, which are carcinogenic in nature are reserved in the finished hosiery supplies and let out as wastewater [11].

Dyes used by the textile industry are now mainly synthetic. They are predominantly extracted from two sources, coal tar and intermediates based on petroleum. Powders, pellets, gels or colloidal dispersions are the state of these dyes in the market. Usually the active ingredient concentrations in the dyes vary from 20 to 80 percent. These new dyes are regularly produced for meeting the demands of new technologies, new kinds of fabrics, detergents, developments in dyeing machineries, along with addressing the significant environmental problems faced by some existing dyes.

With the rapid shift in the textile industry's product profile, from high-cost cotton textiles to durable and flexible synthetic fibres, the pattern of consumption of these dyes is also shifting rapidly. Polyesters now account for a large portion of dye use. Dispersed dyes used in polyesters are therefore also expected to expand at a faster rate.

To make it easier to understand, textile dyestuffs can be grouped into the following groups if we take general dye chemistry as one of the basis for grouping as azoic, acid sulfur, disperse, simple, reactive, mordant, oxidation vat solvent and fluorescent dyes.

Green and Saunders developed one form of coloured azo compound in 1922, in which an amino group is attached to a solubilizing group e.g. methyl sulphate, -CH2-SO3H. In dye bath, they are slowly hydrolyzed and contain azo compound and formaldehyde bi sulphate. This free azo compound was able to color the fibres of cellulose acetate. The name of that dye was "ionamine." This ion amine, however, did not give satisfactory dyeing results. Baddiley and Ellis later developed sulpho ricinoleic acid (SRA) for the dyeing of acetate fibres in 1924. This SRA has been used as an agent for dispersion. SRA was later shown to be capable of dyeing nylon, polyester, acrylic, etc. This dye was called 'Disperse Dye' in 1953. As per the chemical structure disperse dyes are classified as nitro, amino ketone dyes, anthraquinonoid dyes, aono azo dyes, di- azo dyes. In this study we had used azo type of disperse dyes with the commercial names as Coralene Navy RDRLSR, Coralene Red 3G, Rubru RD GLFI dye.

1.3 Background and motivation of the present study

India is the second largest exporter of dyestuffs, after China. It is estimated that approximately 2% of the dyes produced are discharged directly in aqueous effluent, and 10% is subsequently lost during the coloration process [12]. The size of India's market is expected to touch $250 billion in the next two years from $150 billion. Although textile industries play an important role in Indian economy, the waste water management is concerned predominant.

Tiruppur, a textile hub, in Tamil Nadu, has emerged as the leading industrial cluster of cotton knitwear in South India for both the overseas and domestic markets primarily because of climate conditions which facilitate fast processing of yarn. Moreover the availability of raw materials and cheap labour has ensured that the activities of the textile industry here, have experienced rapid growth over the last two decades. Almost 80 percent of India's exports of cotton knitwear are from Tirupur today. There are 6,250 units involved here in different textile industry operations. It consists of 4900 knitting and sewing units, approximately 736 dyeing and bleaching units, 300 printing units, 100 embroidery units and 200 compacting, raising and calendar catering units. Buyers come regularly to visit Tirupur from about 35 countries. In addition to an income matching or exceeding the above amount to cater to the domestic market, this small town contributes around INR.11000 crores (Rs. 110 billion) in foreign exchange earnings annually to the government. In short, Tirupur's economic success depends heavily on this industry and most local people are active in the knitwear sector in one way or another. With a rough estimation of 7×10^5–1×10^6 tons of fabrics produced per year, there are more than 100,000 commercial dyes. Though accurate statistics on the number of dyes released into the surroundings is not known, data shows that 10–15% of the used dyes are released into the environment as they are discharged as wastewater from textile industries causing severe pollution [13].

In Tirupur, the textile industry's water needs have been fulfilled by both surface water and ground water. About 28.8 billion litres of ground water are consumed annually by the units in Tirupur alone. This water is transported in tankers from the surrounding villages such as Avinashi, Palladam, Annur, Kangeyam, and from many parts of the nearby Erode district. Needless to mention, Tirupur faces extreme water shortages. Tankers supply about 80 percent of the total water requirement [14]. The textile units purchase water per tanker at a cost of Rs 250–450. The farmers sell the same thing at Rs 40–80 per tanker. Even at this rate, it is more profitable for farmers to sell water instead of carrying out agricultural activities.

1.4 Major challenges in the treatment of textile effluent

Knowledge of environmental issues has increased dramatically over the past few decades and has become an important concern in textile trade due to different environmental and health regulations, and environmental policy is also increasingly dictated by market forces. Many chemicals used in the textile industry are causing problems with the ecosystem and health [15].

Textile processing is a water intensive process. The waste water generated by the industry is high in BOD, COD, pH, temperature, color, turbidity and toxic chemicals. These polluted effluents ought to be treated chemically to remove the hazardous material and chemicals so that the waste water will comply with the prescribed limits and can be discharged into the public sewer or into aquatic bodies [16]. The major challenges existing in the treatment of textile industrial effluents are recycling and reuse of wastewater, removal of color and reduction of Total Dissolved Solids (TDS) in the treated effluent, generation of huge amount of sludge, large units have invested in Individual Effluent Treatment Units (IETPs) and medium scale units have invested in Common ETPs. However, no techno-economically viable option is available for treating the effluent in small units.

1.5 Existing conventional treatment methods and their limitations

Removal of dye is possible by different or a combination of physical, chemical and biological methods. Adsorption, absorption, membrane filtration, and ultrasonic waves are physical methods; ion exchange, electrolysis, coagulation, traditional and advanced oxidation are chemical methods; and biological methods can be described using algae, fungi, and bacteria [17].

In chemical coagulation, electrostatic gravity between the dye solution and polymeric molecules with opposite loads generates coagulation. The disadvantages of this method are high sludge production and high dissolved solids in treated wastewaters and polymeric molecules with opposite loads generate coagulation. Chemical coagulation is efficient for sulfurous and disperse dyes. Acidic, direct, vat and reactive dyes coagulate with this method too, but do not settle, while cationic dyes do not even coagulate [18]. The conventional treatment methods such as activated sludge, coagulation and flocculation process, reverse osmosis and evaporation consume a lot of chemicals, high energy and are not sufficient to break down azo double bonds of reactive dyes. Also the secondary products or sludge produced becomes more difficult to be disposed. Huge amount of rejects from Reverse Osmosis is another major problem [19].

A large number of well-established traditional decolourisation techniques involving physicochemical, chemical and biological processes, as well as some modern emerging techniques such as sonochemical or advanced oxidation processes, are shown in the available literature. In order to solve this problem, there is no single economically and technically feasible approach and typically two or three methods need to be combined to achieve an acceptable degree of colour removal [20, 21].

The main drawbacks of these physiochemical and biological treatment processes is that they generally lack the broad scope of treatment efficiency required to treat all diverse pollutants present in the textile waste water. The electrochemical technology that was developed can remove chemical components in an effective and economical way.

1.6 Electro chemical methods and their limitations

The electrochemical method is a safer, highly effective treatment method for the treatment of textile liquid waste containing a high concentration of dye. This technique has advantages for decolorization over others, such as the need for simple equipment, better accuracy, and shorter retention time to remove pollutants, less need for chemicals and simpler operation, and [22]. This challenge towards rising demand for quality water can be resolved by the application of certain sophisticated scientific techniques for the conversion of waste water into reusable water. The electrochemical methods for treatment of textile waste water include electro dialysis, electrical ion exchange, electro osmosis, electro oxidation etc. Though these methods have proven to be efficient it is not practically applicable because of its high cost and maintainability [23].

There exists a wide range of methods which can be employed to treat wastewater. With the recent technological development in electrochemical field, a new technique has been introduced in the industry, named as electro-coagulation. EC has been successfully used for decades in order to treat the wastewater of textile, food and protein, phosphate, tannery wastewater, restaurant wastewater and defluoridation. Electro-coagulation among them is cost effective and requires less maintenance. Hence electro-coagulation was applied in this study for the removal of dye from textile wastewater [24, 25].

1.7 Scope of the present study

This study focuses on the treatment of synthetic dyeing wastewater (obtained from Devi Threads, Pachapalayam) using EC method which addresses the need for techno-economically feasible option in treating the effluent from small scale clusters thereby solving the other above mentioned issues of dye removal, high sludge production and reuse of the effluent.

1.8 Objectives of the study

- To prepare TiO_2/Zn from $TiCl_3$ by Spray pyrolysis method and surface characterization of the electrode. (SEM, XRD, EDS)
- To investigate the influence of operational parameters in the removal of dye in textile waste water using TiO_2/Zn in EC process.
- To identify the strategy for improvement of the color removal efficiency (CRE) with reduction in electrolysis time and applied current.
- Sludge characterization.

2. Materials/methods

The detailed methodology of the present study, which was carried out in two phases to address the issues mentioned in the objectives.

- Preparation of TiO_2/Zn Electrode by spray pyrolysis method using the precursor $TiCl_3$
- Removal of disperse dye (Coralene Navy RDRLSR, Coralene Red 3G, Rubru RD GLFI) from synthetic wastewater using TiO_2/Zn Electrode in electrocoagulation process.

2.1 Preparation of TiO_2/Zn electrode by spray pyrolysis

In general the experimental procedure recommended by Beck and Co-workers was adopted for the preparation of TiO_2/Zn electrode by spray pyrolysis of $TiCl_3$ by the following procedure [26]. The TiO_2 coating was done on zinc substrate which was cleaned well and sand blasted to make the surface adherent for the coating. Etching was done for the same purpose using etchants (5% Oxalic acid in water) at 60° C for 1 hour. The surface was then washed properly with water, rinsed using triple distilled water.

Then it was coated with the precursor solution $TiCl_3$ (0.05 N) 4 ml, 1:1 HNO_3 water (2.5 ml). The precursor solution was cooled to 5°C to which isopropyl alcohol (15 ml) was added maintaining the temperature 5 to 10°C.

2.2 Deposition of TiO_2 coating on zinc by spray pyrolysis

Spray pyrolysis technique used in this work is a tubular reactor type as it is given in **Figure 1**. The precursor solution is pumped through a 100 kHz ultrasonic atomizer (Lechler Inc.) to the heated substrate held within a tubular quartz reactor (5 cm in i.d. and 30 cm in length). On the graphite sample holder, the zinc metal substrate was placed and heated by quartz heating bulbs located on top and below the tube. The temperature of the substrate was controlled by a thermocouple type K connected to the temperature controller display. The solution was pumped into the atomizer once the temperature controller indicated the desired temperature, and the atomized droplets of the precursor were dispersed and transported using nitrogen (N_2) and oxygen (O_2) as the carrier gases towards the substrate to the reaction chamber. The aerosol droplets were decomposed and transformed to very tiny

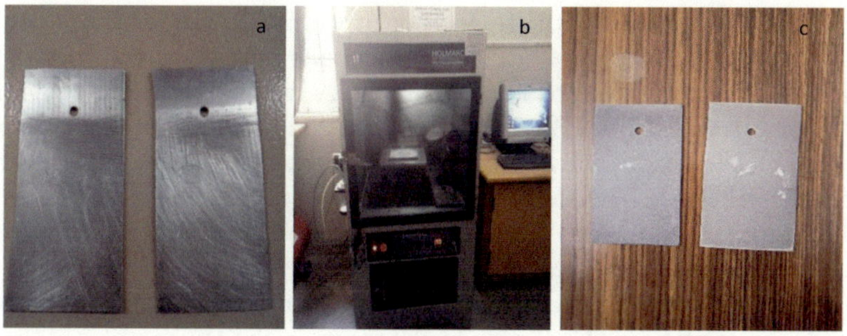

Figure 1.
(a) Photograph of the pretreated zinc plate, (b) Coating on zinc in spray pyrolysis, (c) TiO_2/Zn.

particles of oxide compounds upon meeting the hot surface of the zinc substrate. The usual spray time for one time deposition of a 50 ml solution batch was around 30–45 min depending on the spray conditions. The sample was then cooled down by 5–10°C/min from the deposition temperature slowly to room temperature. The electrodes were prepared with dimensions of 9 cm × 3.2 cm × .5 cm. **Figure 1** shows the photograph of the pretreated zinc plate, coating on zinc in spray pyrolysis and the TiO_2/Zn plate.

2.3 Electrocoagulation process

Disperse dye was obtained from one of the textile industry in Coimbatore, Tamil Nadu. (Commercial name; Coralene Navy RDRLSR, Coralene Red 3G, Rubru RD GLFI) The simulated wastewater was prepared by dissolving a 150 ppm of disperse dyes in distilled water. The experimental device is schematically shown in **Figure 2**. The EC unit consists of an electrochemical reactor which is a glass beaker with magnetic stirring, a D.C power supply and two sets of EC process were run with TiO_2/Zn – TiO_2/Zn and Zn – Zn. (Hereafter these electrodes will be represented as 'A' (TiO_2/Zn – TiO_2/Zn) and 'B' (Zn – Zn)).

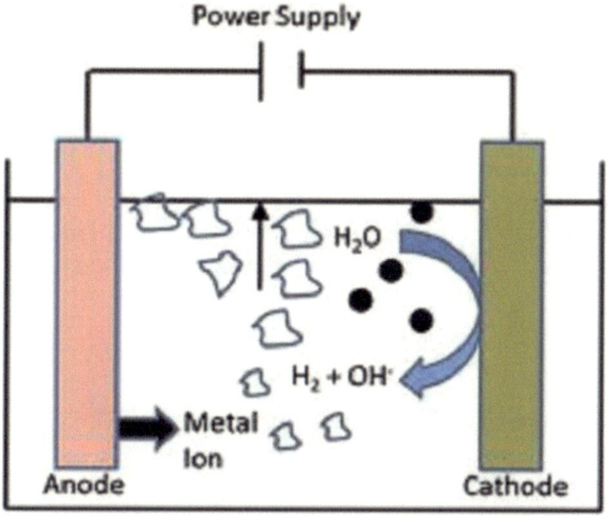

Figure 2.
Schematic representation of EC process (Source: http://pubs.sciepub.com).

The electrodes were used with dimensions of 9 cm × 3.2 cm × .5 cm. The total effective electrode area was 28.16 cm^2 and the spacing between electrodes was 1 cm. The electrodes were connected to a digital dc power supply (var tech) providing a current ranging from 0.05 to 0.3A. 500 ml electro coagulation cell that contained the 250 ml test solution and a magnetic stirrer was used to stir the solution, thereby enhancing the efficiency. The applied current was adjusted to a desired value and the coagulation was started. In each run, 250 ml of dye solution was placed into the electrolytic cell. Before each run, electrodes were washed with water and dipped in 15% hydrochloric acid in order to remove dust from the electrode plates and thus weigh the electrodes after drying. The electrode plates are washed with water at the end of each run, dried and weighed at last. Whattman filter paper was used to filter the subsequent treated sample and filtrate was used for the analysis. The individual effects of electrolysis time and applied current on colour removal efficiency were quantified in this analysis (CRE). The pH was adjusted by adding 0.5 M HCl or 0.5 M NaOH. The conductivity of solutions was raised and adjusted to different values by the addition of NaCl. All experiments were carried out at constant temperature of 25°C. Two sets of EC process were carried out with 'A' and 'B' electrodes with operating parameters like pH, EC time, dye concentration, addition of electrolyte NaCl and applied current for optimization to achieve higher Color Removal Efficiency (CRE%). On the basis of the initial experiments, other parameter rates were considered constant. The experimental set up in EC process, after the experiment, the coagulated dye solution getting separated as sludge settling and floating are given in **Figure 3**.

2.4 Colour removal efficiency

The ultraviolet–visible spectrophotometer (Jasco V-670 spectrophotometer) was used to measure the wavelength (542 nm) of dye. The calculation of color removal efficiencies (CRE %) after electrocoagulation treatment was performed using the formula, as it is given in Eq. (1).

$$\text{Colour removal efficiency}\,(\text{CRE\%}) = 100 * (A_i - A_f) / A_i \qquad (1)$$

Where A_i and A_f were the absorbance of the dye in solution afore electrocoagulation and at the time *t*, respectively. Using a conductivity metre (Elico CM180) and a pH metre (Eutech), the electrical conductivity and pH of different dye concentrations were measured.

2.5 Sludge characterization

EC method is said to generate less sludge compared to chemical coagulation, eventhen it is important to take appropriate measures for the reuse or safe disposal

Figure 3.
Photograph of the (a) dye solution, (b) after EC sludge settling, (c) sludge floating.

of sludge in order to determine the quantification and characterization of the sludge generated and characterization of the sludge plays a major role in deciding healthy discarding or reclamation. Therefore by filtration technique, the sludge formed during the process was separated and placed in the drying oven for about 24 hours at 100°C and dried, which was then weighed [27].

3. Characterization techniques

3.1 Adhesion test

The adhesion strength between the coating and the substrate was tested by applying the Scotch tape test on the deposited films before any characterization. In general, if the film adheres to the substrate and it does not peel off the substrate, adhesion strength is considered to be good and it was found good for TiO_2/Zn.

3.2 Microstructural and phase characterization

The morphological, and structural analysis of TiO_2/Zn electrodes were carried out by X-Ray diffractometer and Scanning Electron Microscope (SEM).

Non-destructive tool X-ray diffraction (XRD) is used for the identification and determination of structural properties such as train, grain size, expitaxy, phase composition, and crystalline phases and orientation of the deposited film. The research was conducted to study the micro structure of the particles present in the electrode and sludge using the JEOL-JDX 8030 Model 6000 diffractometer with Cu-K a radiation (λ = 0,15406 nm).

Surface morphology studies were performed by JEOL, JSM 35 CF, Japan, using a Scanning Electron Microscope (SEM). Energy Dispersive X-Ray Spectroscopy Energy Dispersive X-Ray Spectroscopy (EDS) was used for the elemental analysis.

4. Results and discussion

4.1 Characterization of TiO_2/Zn

The surface morphology, composition of elements present and the microscopic structure in the electrode were analyzed using SEM, EDS and XRD.

Zinc substrate and TiO_2 coated zinc by spray pyrolysis were studied for their surface morphology and elemental analysis using Scanning Electron Microscope (SEM) and Energy Dispersive X-Ray Spectroscopy. The SEM micro-graph and EDS spectrum of zinc and TiO_2/Zn are given in **Figure 4**.

The TiO_2 SEM micrograph showed dense particles and their shape was in regular pattern. Obviously, the Ti and O peaks can be seen in both EDS spectra with peaks with chlorine on the coated surface. Morphology of SEM monographs show that the surface are porous. Furthermore the cracks in the surface layer are evident beyond the pores, which may imply too fast volume growth of the TiO_2 layer on the zinc substrate.

The elemental constituents of TiO_2 nanoparticles were analysed using EDS, **Figure 4** shows the EDS spectra for TiO_2 particles on the zinc substrate, peaks around 0.2, 0.3 and 4.3 keV are attributed to the binding energies of titanium and oxygen of TiO_2. This finding confirms the existence of elemental compounds with an impurity peak corresponding to chlorine from the precursor $TiCl_3$ on the zinc substrate. The peaks of 0.25, 8.5, 9.8 keV are linked to the oxygen-binding energies of zinc.

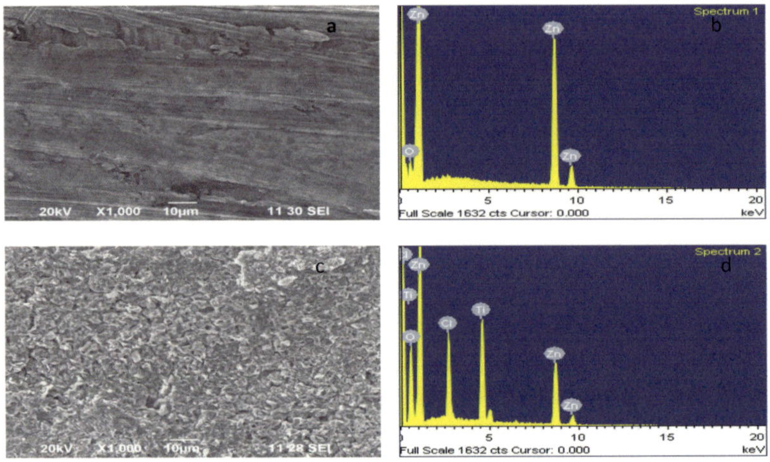

Figure 4.
(a, b) SEM, EDS image of Zinc, (c, d) TiO_2/Zn.

The X-ray diffractograms of TiO_2/Zn and zinc are shown in **Figure 5**. The diffraction peaks of TiO_2 phase, and the substrate zinc are present in the investigated electrode.

The particle size of TiO_2 is related to the diffraction peak broadening, so X-ray diffraction spectra of coated TiO_2 nanoparticles were taken and particle size and phase composition were determined. The lattice parameter observed a = b = 3.780, c = 9.513. The average particle size calculated by using Scherer Eq. indicated high surface area [28].

The particle size of nanomaterial is related to the diffraction peak broadening, so X-ray diffraction spectra of synthesized TiO_2 nanoparticles were taken and peak size and composition were determined. Sharp peaks obtained corresponding to the planes (104), (018), (110), (024), (024) and (300) confirmed the nanocrystalline anatase structure. It shows the primitive hexagonal structure of nanoparticles of TiO2. TiO2 deposition is consistent with 2 theta values (30,32,35,47,56,63,68) and Zn deposition with 2 theta values (36,43,54,70,78) from the XRD results. The data was compared with JCPDS card no:71-1059 for TiO_2 and 65-3358 for Zn. In the XRD pattern, no other impurity peak was observed.

4.2 Optimization of operational parameters

Optimization of operational parameters in electrocoagulation process such as (applied current, pH, reaction time, NaCl (electrolyte) concentration) were investigated. Optimization of the above mentioned operational parameters is necessary to improve the performance of EC process and the economic viability. Since the influence of these parameters depend on the type of waste water and its concentration, optimal value of the operational parameters were identified for the textile dyeing waste water synthesized using disperse dye powder. The CRE at varied current supplied for the process, pH, reaction time, and the concentration of the electrolyte used is discussed here with. Two sets of EC with TiO2/Zn-TiO2/Zn (electrode 'A') and Zn-Zn (electrode 'B') were conducted for optimization and the results were compared.

4.3 Effect of pH

The pH is one of the most important parameters in the performances of EC process in order to achieve a compromise between best coagulation and best

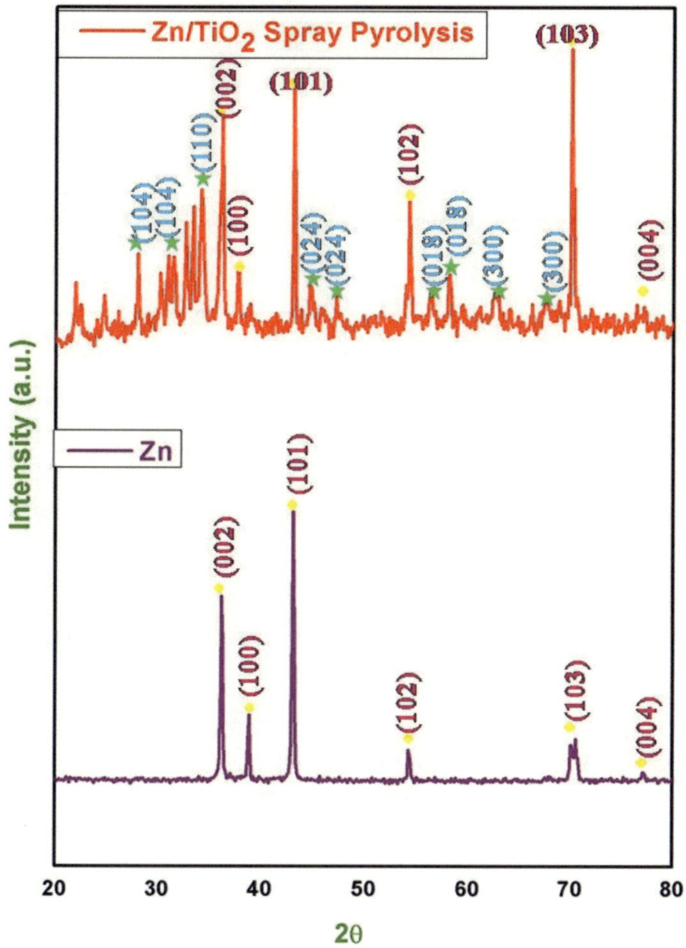

Figure 5.
XRD Pattern for Zn and TiO_2/Zn.

flotation. The optimum range may however vary as a function of electrode material and dye structure. And the fact is that pH determinates the metallic ions speciation, the chemical state of other species in the solution, and solubility. Hence the optimum pH is necessary to minimize environmental remediation costs and make the process more efficient. To investigate this effect in this work, a series of experiments were performed using synthetic disperse dye solution. Experiments were carried out at various values of pH (4.5, 5.5, 6.6, 7.5, 8.5 and 9.5) under which the applied current was kept at 0.05A for electrode 'A' and 0.15A for electrode 'B'. The pH was adjusted to a desirable value using NaOH, or H_2SO_4 and varied in the range 4.5 to 9.5. The treated sample is collected and filtered and the % of CRE estimated. The obtained results are shown in **Figure 6(a)**.

It can be noticed that decolorization was most effective in a pH range between 4.5 to 8.5 for coated and 4.5 for zinc electrodes and removal of CRE% reached values between 91.7 to 98.1 for the coated and 80.4 for Zn electrodes. This refers to the area where $Ti(OH)^{2+}$ and $Ti_2(OH)^{2+}$ would have been formed, and $Ti(OH)_3$ as insoluble species prevailed. On the other hand, above pH 8.5, % of CRE fell when soluble $Ti(OH)_4^-$ anions become predominant at high pH. At an initial pH of 4.5, dye process was completed within 5 minutes with high efficiency. And also at pH 4.5, the dye separated well from solution and the sludge floated with good

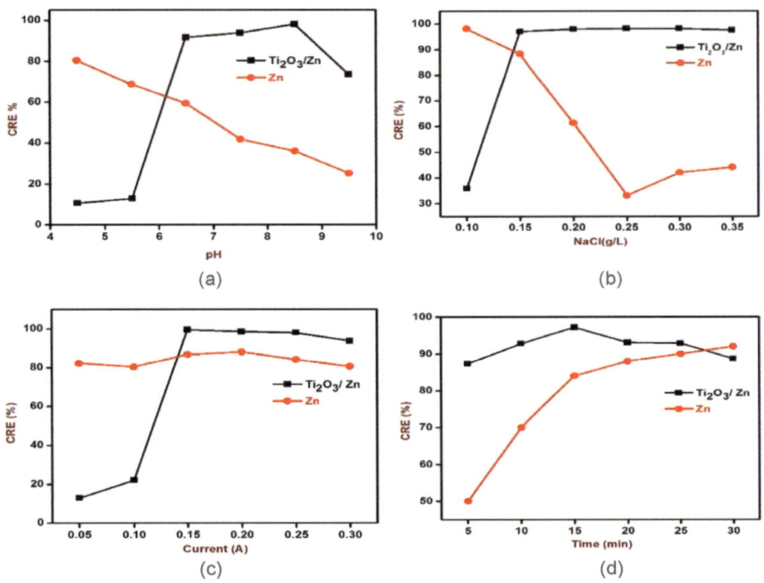

Figure 6.
Effect of varying (a) pH on CRE (%), (b) concentration of NaCl on CRE (%), (c) applied current on CRE (%), (d) Time on CRE (%).

percent of CRE removal in the case of electrode 'A', and the separation was clear and high % of CRE removal as the pH was increased from 4.5 to 7.5 may be due to the formation of more monomeric and polymeric insoluble Ti species.

The percentage of CRE was 80.49 in the case of zinc electrodes at pH 4.5, as the pH increased from 4.5 to 9.5 there was a gradual and sudden decrease in percent of CRE, this could be attributable to the development of more soluble Zn(OH)2 than insoluble zinc ion and prevented coagulation.

4.4 Effect of electrolyte

NaCl is typically used by electrocoagulation treatment to improve the solution conductivity, so that current consumption could be reduced. In addition, increasing water conductivity using NaCl has other advantages; for example, chloride anions can significantly reduce the adverse effects of other anions, such as bicarbonate and sulphate ions. Conversely, an excessive amount of NaCl induces an overconsumption of the TiO_2/Zn electrodes due to 'corrosion pitting'; zinc dissolution may become irregular. This is the reason why NaCl addition should be limited and optimized. It also allows the passivity of the electrodes to be decreased by removing the surface passivation of the oxide film produced because of its catalytic action on the electrode surface. The adverse effects of other anions (due to their oxidation) and the availability of metal hydroxide in the solution could greatly reduce chloride ions. This parameter could, therefore have a substantial effect on the efficiency of pollutant removal. The ability to remove pollutants under certain conditions depends on the amount of coagulant produced that is related to the media's conductivity. In order to improve the conductivity of the waste water to be treated, NaCl is applied in this process and the increase in salt concentration increases the concentration of ions in the solution, thus reducing the electrical resistance of the solution, thus reducing the resistance between electrodes. Greater the electrical conductivity means higher electrical conductivity of the solution, and for creating a fixed current conductivity, a higher voltage is needed. Therefore, for creating a fixed current conductivity, less NaCl is needed compared with other electrolytes like Na_2SO_4, or Na_2SO_3, Moreover NaCl has a higher ionization speed and mobility due to the lower radiuses of Na and Cl as a result, and more current passes through wastewater and by increasing passing current, the speed of anode dissolution increases. On the other hand, producing acidic species such as HCl and ClO^- enhances the desirability of revival conditions. Therefore, using NaCl as electrolyte has advantage for lower price. Also, textile and dying industries use plenty of NaCl and wastewaters of these industries comprise ions of NaCl, because it is cheap and the solution containing this salt has high conductivity thus it need low voltage for electro-coagulation and so it is economical in industrial scale.

To study the effect of wastewater conductivity on dye removal, various experiments were performed using NaCl as the electrolyte in the range of 0.1–0.35 g/L and the CRE removal efficiencies observed during EC process are given in **Figure 6(b)**.

It is believed that the main pollutant removal mechanism observed during electro-coagulation is adsorption and entrapment onto the amorphous titanium and $Zn(OH)_2$ precipitate formed due to the anodic reaction at maximum rate at pH 8.5 and 4.5 for 'A' and 'B' electrodes and the anodization of Ti, Zn are given in Eqs (2)and (3)

$$Ti(s) \rightarrow Ti^{3+}(aq) + 3e^- \quad (2)$$

$$Zn \rightarrow Zn^{2+}(aq) + 2e^- \quad (3)$$

As is evident in **Figure 6(b)** increasing the electrolyte concentration from 0.1 to 0.35 g/L the high % of CRE reducing rates (98.2 for electrode 'A' and for 'B' 98.1), Due to the improved conductivity of the aqueous medium and the addition of NaCl up to 0.35 g/L, the improvement in the removal efficiency of CRE can be related to a shift in ionic strength resulting in a moderate but substantial decrease in treatment efficiencies in terms of CRE removal in the electrode 'A.' But there was a sudden decrease in the CRE in the case of Zn electrodes as the electrolyte concentration increased from 0.1 g/L (from 98 percent sudden decrease in percentage of CRE) a further increase in electrolyte concentration did not improve, these findings can be clarified by the fact that when the NaCl concentration increased, the electrolyte conductivity correspondingly increased. This was possibly because the chloride ions could destroy the passivation layer and increase the metal's anodic dissolution rate, either by integrating chloride ion into the oxide film or by involving the same in the electrochemical reaction. A further increase in NaCl concentration in both electrodes showed negative degradation and salt film formation on the electrode surface, which would obstruct the interaction between the electrode and the waste water. The chance of successful interaction between the organic contaminants and the hydroxyl radicals was therefore reduced.

4.5 Effect of applied current

It is clear that the applied current is strongly influenced in EC. Increased current results in increased anodic dissolution of metal ions, resulting in the formation of high quantities of precipitate for pollutant removal. In batch electro-coagulation, operating current density is important as it is the only operational parameter which can be directly regulated. The quantity of oxidised metal ions increased when the current increased, and the quantity of metal hydroxide compounds for pollutant precipitation and adsorption also increased. In addition, as the applied current grows, the rate of development of hydrogen bubbles increases and their sizes decrease. For successful dye removal, all these effects are significant. Other side reactions in the vicinity of the anode, such as the direct oxidation of one of the components of the contaminant or the formation of oxygen that restricts the efficacy of electro-coagulation, can be caused by high-current activity. In comparison, a high current causes the passivity of the cathode to decrease, resulting in high energy consumption. The best conditions correspond to a low applied current and substantial electrolysis time especially in terms of energy consumption and electrode consumption. It is essentially critical to avoid operating at too high current to address the excessive generation of Ti, Zn poly hydroxides in wastewater. The efficiency of the reduction of contaminants depends on the production of Ti (IV or III) and Zn (II) by the anode, so that a high period of electrolysis will lead to a higher production of titanium hydroxide or zinc hydroxide, which is responsible for the coagulation process. To optimise the applied current in the EC method, experiments were performed under other optimised parameters at different applied current from 0.05 to 0.30A and the results are given in **Figure 6(c)**.

As the applied current increased from 0.05 to 0.30A, the anodisation accompanied by floc formation with the Ti^{n+} ions increased, the CRE percentage was 99.5 at 0.15A and consequently the removal efficiency decreased marginally in the case of electrode 'A' and in Zn-Zn electrodes as the applied current increased, the CRE removal increased gradually, the maximum CRE was 86 percent at 0.15A. The increase using the electrode 'A' was due to the dissolution of both Ti^{3+} and Zn^{2+}, whereby at lower applied current highest CRE removal was obtained, moreover it could be observed that increasing the applied current above the optimum value decreases the CRE is due to undesirable side reaction such as electrolysis of water and oxygen evolution from OH free radicals. However energy consumption leads to be higher for increased applied current and indicates that increase in current density led to less efficient sprocess.

4.6 Effect of time

The percentage color removal efficiency depends directly on the concentration of ions produced by the electrodes. This can be achieved by parameter like EC time. Because the formation of metal ions and concentration of the metal hydroxides play an important role on pollutant removal, this depends on operation time. To study its effect, the EC time was varied for different time intervals i.e., 5, 10, 15, 20, 25, 30 min and the other optimized parameters were kept constant. The results obtained are illustrated in **Figure 6(d)**. The H_2 and O_2 release and flocs formation increased over time and the foam became thicker.

A plot is drawn between time verses % CRE for two different electrodes as shown in graph, it was clearly known that as the time increases the percentage of CRE increased in the case of electrode 'A' at 15 mins was 97.2 and then there was slight reduction in CRE. Whereas in the case of electrodes 'B' there was a gradual increase of CRE from 50 to 92% and high % of CRE was achieved at 30 minutes which will be with more of energy as well electrode consumption compared to the electrodes 'A'.

In the EC process, the anode produces metal ions during electrochemical reaction. Metal ions are destabilization agent. If the charge loading were low, the metal ion released from the anode would not be sufficient to destabilize all the colloidal and suspended particles, so dye removal was not efficient in the case of zinc electrodes at applied current 5A. When EC time changed from 5 to 30 minutes the energy consumption increased from 0.0004966 to 0.00174 kWh/m^3in the case of electrode 'A' and 0.000843 to 0.005227 kWh/m^3 forZn electrodes. From these results it is shown that % CRE removal is high with less energy consumption in the case TiO_2/Zn than zinc electrodes. Treatment time is related with energy

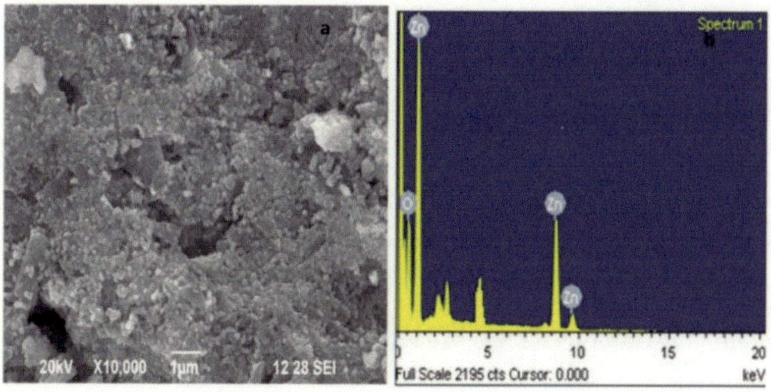

Figure 7.
a) SEM, b) EDS of Sludge after electrocoagulation.

consumption and wastewater treatment performance. It is well known that the removal efficiency did not improve much after 25 min electrolysis, but the prolonged time would increase the electrochemical treatment cost. The results indicate that the optimum electrolysis time for best removal efficiencies is 10 min for TiO_2/Zn and 30 min for Zn electrodes.

4.7 Sludge charaterization

The SEM images that display the sludge's morphology characteristics are shown in **Figure 7**. It can be visualised that the flaky structure of the particulates produced from electrocoagulation is confirmed by the adsorbed dye molecules (grey or black portion) on the surface and the presence of a peak at 4.8 keV in the EDS spectra confirms the titanium ion floc. It is assumed from SEM, EDS sludge study, that the first titanium flocs which were involved in electrocoagulation, once after the depletion of the same, then the dye molecules have been removed by zinc flocs.

5. Conclusion

The results obtained in this study shows that the color can be eliminated with high percentages using the newly prepared TiO_2/Zn electrodes from $TiCl_3$ by spray pyrolysis method compared with using Zn in EC process.

The effect of applied current on CRE% can clearly be understood. As the applied current increased the removal of organic matter also increased. In all the experimental parameters, dye removal efficiencies (CRE%) was observed in short period of operational time in using TiO_2/Zn. The increase in applied current is also considered to mean an increase in energy usage. Thus, due to local discharge limits, energy and electrode usage, local energy unit prices and some other limiting factors, an acceptable current density has to be considered.

Furthermore this research showed that the initial pH of the waste water at which high CRE% was obtained as the optimized pH 8.5 in the case of TiO_2/Zn but 4.5 for Zn electrode. Overall, high color removal efficiency of disperse dye was obtained using newly modified TiO_2/Zn.

The EC process has the potential to treat the textile dyeing wastewater and thus to reduce the contamination of the environment by the dye molecules as the real time textile dyeing waste water can be treated with this newly developed TiO_2/Zn. Further work can be done as given.

- Preparation of TiO_2/M electrode.
    - Attempt can be made in the preparation of TiO_2/M (M for metal) with different metals as electrode materials,
    - different precursor
    - preparation parameters like flow rate, number of coating etc. can be adopted in spary pyrolysis.
- Opretaional parameters in EC process
    - In EC process the other operational parameters like varying the distance between the electrodes, height of electrodes immersed in the waste water etc. can be optimized.
    - Investigation other than CRE%, like COD, BOD etc. can be calculated.
- Economic parameters and opretional cost can be designed for the other opretaional parameters like
    - Applied current,
    - dye concentration,
    - electrolyte concentration
    - and pH,
- The electrode can be applied to dyes other than disperse dyes and the real time waste water in dyeing units.

Author details

Parameswari Kalivel
Department of Chemistry, Karunya Institute of Technology and Sciences, Coimbatore, India

*Address all correspondence to: parameswari@karunya.edu

References

[1] Verma, A.K., Dash, R.R. and Bhunia, P, A Review on Chemical coagulation/Flocculation Technologies for Removal of Colour from Textile Wastewaters. Journal of Environmental Management, 2012; 93:154-168. doi.org/10.1016/j.jenvman.2011.09.012

[2] Farré M la, Pérez S, Kantiani L, Fate and toxicity of emerging pollutants, their metabolites and transformation products in the aquatic environment, TrAC Trends in Analytical Chemistry, 2008; 27:991-1007. doi: 10.1016/J.TRAC.2008.09.010

[3] Kivaisi AK. The potential for constructed wetlands for wastewater treatment and reuse in developing countries: a review. Ecological Engineering, 2001; 16:545-560. doi: 10.1016/S0925-8574(00)00113-0.

[4] Shannon MA, Bohn PW, Elimelech M, et al (2009) Science and technology for water purification in the coming decades. In: Nanoscience and Technology. Co-Published with Macmillan Publishers Ltd, UK, 2009; 337-346.

[5] Verma, A.K., et al., A review on chemical coagulation/flocculation technologies for removal of colour from textile wastewaters. J. Environ Manag. 2012; 93: 154-168. doi: 10.1016/j.jenvman.2011.09.012

[6] Chowdhury, S., Mishra, R., Saha, P. and Kushwaha, P. 2011. Adsorption thermodynamic kinetics and Isosteric heat of adsorption of Malachite Green onto chemically modified rice husk. Desalination. 2011;265: 159-168. *doi*:10.1016/j.desal.2010.07.047

[7] Gupta, V.K., Mittal, A., Gajbe, V. and Mittal, J. 2008. Adsorption of basic fuchsin using waste materials bottom ash and deoiled soya—as adsorbents. J Colloid Interface Sci. 2008; 319: 30-39. doi: 10.1016/j.jcis.2007.09.091

[8] Anjaneyulu Y, Sreedhara Chary N, Samuel Suman Raj D, Decolourization of Industrial Effluents – Available Methods and Emerging Technologies – A Review. Reviews in Environmental Science and Bio/Technology, 2005; 4: 245-273. doi: 10.1007/s11157-005-1246-z.

[9] Robinson T, McMullan G, Marchant R, Nigam P, Remediation of dyes in textile effluent: a critical review on current treatment technologies with a proposed alternative. Bioresource Technology, 2001; 77:247-255. doi: 10.1016/S0960-8524(00)00080-8.

[10] Sarayu K, Sandhya S, Current Technologies for Biological Treatment of Textile Wastewater–A Review. Applied Biochemistry and Biotechnology, 2012;167:645-661. doi: 10.1007/s12010-012-9716-6.

[11] Haseena M, Malik MF, Javed A, et al. Water pollution and human health. Environmental Risk Assessment and Remediation, 2017;1:3 doi: 10.4066/2529-8046.100020

[12] Garg, V.K. and Kaushik, P. 2008. Influence of textile mill wastewater irrigation on the growth of sorgum cultivars. Appl Ecol and Enviro Res. 2008; 6: 1-12.

[13] Gupta VK, Suhas, Application of low-cost adsorbents for dye removal – A review. Journal of Environmental Management, 2009;90:2313-2342. doi: 10.1016/J.JENVMAN.2008.11.017.

[14] Sarathi JN, Rao SK, Mba VK Environmental issues and its impacts associated with the textile processing units in Tiruppur, Tamilnadu.

[15] Forgacs E, Cserháti T, Oros G, Removal of synthetic dyes from wastewaters: a review. Environment International, 2004; 30:953-971. doi: 10.1016/J.ENVINT.2004.02.001.

[16] Arslan, A., et al., Effect of Operating Parameters on the Electrocoagulation of Simulated Acid Dyebath Effluent the Open Environmental & Biological Monitoring Journal, vol. 1. Bentham Science Publishers Ltd., 208; 1-7. doi: 10.2174/1875040000801010001

[17] Saha, G., et al., 2015. A low cost approach to synthesize sand like AlOOH nanoarchitecture (SANA) and its application in defluoridation of water. J. Environ. Chem. Eng. 2015; 3: 1303-1311. Doi.org/10.1016/j.jece.2014.11.030.

[18] Ahangarnokolaei MA, Ganjidoust H, Ayati B (2018) Optimization of parameters of electrocoagulation/flotation process for removal of Acid Red 14 with mesh stainless steel electrodes. Journal of Water Reuse and Desalination, 2018;8:278-292. doi: 10.2166/wrd.2017.091.

[19] Ghaly A, Ananthashankar R, Alhattab M, Ramakrishnan V, Journal of chemical engineering & process technology. OMICS Publishing Group, 2014; 5:1-18 DOI: 10.4172/2157-7048.1000182

[20] Muga HE, Mihelcic JR, Sustainability of wastewater treatment technologies. Journal of Environmental Management, 2008; 88:437-447. doi: 10.1016/J.JENVMAN.2007.03.008.

[21] Chen G, Electrochemical technologies in wastewater treatment. Separation and Purification Technology, 2004; 38:11-41. doi: 10.1016/J.SEPPUR.2003.10.006.

[22] Lin SH, Peng CF, Treatment of textile wastewater by electrochemical method. Water Research, 1994; 28:277-282. doi: 10.1016/0043-1354(94)90264-X.

[23] Anglada angela, Urtiaga A, Ortiz I, Contributions of electrochemical oxidation to waste-water treatment: fundamentals and review of applications. Journal of Chemical Technology & Biotechnology, 2009; 84:1747-1755. doi: 10.1002/jctb.2214.

[24] İrdemez Ş, Demircioğlu N, Yildiz YŞ,The effects of pH on phosphate removal from wastewater by electrocoagulation with iron plate electrodes. Journal of Hazardous Materials, 2006; 137:1231-1235. doi: 10.1016/J.JHAZMAT.2006.04.019.

[25] Emamjomeh MM, Sivakumar M, Review of pollutants removed by electrocoagulation and electrocoagulation/flotation processes. Journal of Environmental Management, 2009; 90:1663-1679. doi: 10.1016/J.JENVMAN.2008.12.011.

[26] F.Beck and W. Gabriel, *Angew. Chem. Int. Ed. Engl.*, 1985; 24:771, doi.org/10.1002/anie.198507711

[27] Parameswari Kalivel, Rajkumar Pluto Singh, et al., Elucidation of Electrocoagulation Mechanism in the Removal of Blue SI dye from aqueous solution using Al-Al, Cu-Cu Electrodes - A Comparative Study" Ecotoxicology and Environmental Safety, 2020; 201: 110858. doi.org/10.1016/j.ecoenv.2020.110858.

[28] Parameswari Kalivel, Jegathambal Palanichamy and Mano Magdalene Rubella Potential of Ti_2O_3/Zn Electrodes versus Zn by Electrocoagulation Process for Disperse Dye Removal Asian Journal of Chemistry; 2019; 31:8, 1835-1841. https://doi.org/10.14233/ajchem.2019.22097.